SCALE

Geographical scale is a central concept enabling us to make sense of the world we inhabit. Among other things, it allows us to declare one event or process national and another global or regional. However, geographical scales and how we think about them are profoundly contested, and the spatial resolution at which social processes take place (local, regional, or global) together with how we talk about them, has significant implications for understanding our world.

Scale provides a structured investigation of the debates concerning the concept of scale and how various geographical scales have been thought about within critical social theory. Specifically, the author examines how the scales of the body, the urban, the regional, the national, and the global have been conceptualized within Geography and the social sciences more broadly. The first part of the book provides a comprehensive overview of how different theoretical perspectives have regarded scale, especially debates over whether scales are real things or merely mental contrivances and/or logical devices with which to think, as well as the consequences of thinking of them in areal versus in networked terms. The subsequent five chapters of the book each take a particular scale – the body, the urban, the regional, the national, the global – and explore how it has been conceptualized and represented discursively for political and other purposes. A brief conclusion draws the book together by posing a number of questions about scale which emerge from the foregoing discussion.

The first single-author volume ever written on the subject of geographical scale, this book provides a unique overview by pushing understandings of scale in new and original directions. The accessible text is complemented by didactic boxes, and *Scale* serves as a valuable pedagogical reference for undergraduate and postgraduate audiences wishing to become familiar with such theoretical issues.

Andrew Herod is Professor of Geography and Adjunct Professor of International Relations and of Anthropology at the University of Georgia. He is also an elected official, serving in the government of Athens-Clarke County, Georgia.

Key Ideas in Geography

SERIES EDITORS: SARAH HOLLOWAY, LOUGHBOROUGH UNIVERSITY AND GILL VALENTINE, SHEFFIELD UNIVERSITY

The *Key Ideas in Geography* series will provide strong, original, and accessible texts on important spatial concepts for academics and students working in the fields of geography, sociology and anthropology, as well as the interdisciplinary fields of urban and rural studies, development and cultural studies. Each text will locate a key idea within its traditions of thought, provide grounds for understanding its various usages and meanings, and offer critical discussion of the contribution of relevant authors and thinkers.

Published

Nature
NOEL CASTREE

City
PHIL HUBBARD

Home
ALISON BLUNT AND ROBYN DOWLING

Landscape
JOHN WYLIE

Mobility
PETER ADEY

Migration
MICHAEL SAMERS

Scale
ANDREW HEROD

Forthcoming

Rural
MICHAEL WOODS

SCALE

Andrew Herod

Routledge
Taylor & Francis Group

LONDON AND NEW YORK

First published 2011 by Routledge
2 Park Square, Milton Park, Abingdon, Oxon, OX14 4RN

Simultaneously published in the USA and Canada
by Routledge
605 Third Avenue, New York, NY 10017

Routledge is an imprint of the Taylor & Francis Group, an informa business

Copyright © 2011 Andrew Herod

Typeset in Joanna and Scala Sans by
Florence Production Ltd, Stoodleigh, Devon

British Library Cataloguing in Publication Data
A catalogue record for this book is available from the British Library

Library of Congress Cataloguing in Publication Data
Herod, Andrew, 1964–.
   Scale/by Andrew Herod.
      p. cm. – (Key ideas in geography)
   Includes bibliographical references and index.
   1. Geographical perception.   2. Map scales.   3. Urban
   geography.   4. Cities and towns.   I. Title.
   G71.5.H47 2010
   912.01'48 – dc22   2010005089

ISBN13: 978-0-415-34907-9 (hbk)
ISBN13: 978-0-415-34908-6 (pbk)

CONTENTS

LIST OF ILLUSTRATIONS

FIGURES

TABLES

BOXES

ACKNOWLEDGMENTS

Parts of Chapter 1 draw upon Herod and Wright (2002a) and Herod (2008). Chapter 6 draws upon Herod (2009, particularly Chapters 2 and 4). The image in Box 2.1 is courtesy of The M.C. Escher Company-Holland. The image in Box 6.1 is courtesy of Deutsches Historisches Museum, Berlin, Germany/© DHM/The Bridgeman Art Library.

PREFACE

Of all the concepts that geographers (and others) use to understand the world around them, scale is a – or perhaps even the – central one. Indeed, Howitt (1998: 50) has called scale one of Geography's "foundational concepts."[1] Thus, the concept of scale enables us to talk about matters of spatially uneven development, wherein the spaces within a particular scalar unit (such as a region) are seen to be more similar than those which exist beyond it, or to name certain processes "global" ones and others merely "local" ones. Hence, understanding the world as scaled provides us with a sense of size (as when "the local" is seen as smaller than "the global"), of power relationships (as when "the global" is thought to be undermining the influence of other scales such as "the national" and "the local"), and, often, of hierarchy (as when we consider the global scale more important than scales such as the national). At the same time, however, scale is a complex concept, for two reasons. First, it has at least two very different meanings. On the one hand, there is a quite technical meaning, one referring to the ratio between the size of objects on the Earth's surface and their size when represented on a map – as when a cartographer states that a map has a scale of, say, "one to one million," with one inch on the map representing one million inches (approximately 15.8 miles) on the Earth's surface. On the other hand, the term "scale" is used as a kind of shorthand to describe either an areal unit on the Earth's surface (as when studying a phenomenon "at the regional scale") or the extent of a process's or a phenomenon's geographical reach (as when suggesting that a particular process is "a regional" or "a national" one). Second, there is no clear agreement on whether scales "actually exist" – that is to say, there is no clear agreement as to whether scales are real things, materially manifested in the landscape, or whether they are simply mental devices by which we categorize and make sense of the world. Their ontological status, in other words, is the subject of much debate.

Although scale has always been a central concept in geographical and other thinking, during the past two decades or so a veritable cottage industry of critical writings and ponderings about it has emerged. Much of this has been related to contemporary economic, political and cultural restructurings – together with new ways of thinking about such restructurings – associated with what has come to be called "globalization" (Herod 2009). Thus, a central theme in much globalization writing has seen the process as one whereby scales such as the national or the local are being eviscerated by the global, with globalization bringing about the "delocalization" (Virilio 1997) and/or "denationalization" (Sassen 2003) of economic and political life. Hence Gray (1998: 57) has argued that behind all the different conceptualizations of globalization (is it good or bad? is it an economic, a political, or a cultural process, or all three? is it inevitable or not?) "is a single underlying idea which can be called *de-localization*: the uprooting of activities and relationships from local origins and cultures. [Globalization] means the displacement of activities that until recently were local into networks of relationships whose reach is distant or worldwide." This has raised thorny questions. For instance, what does it really mean when we say that what started as a "local" family business has now grown to become a "national" or a "global" transnational corporation? Does it make a difference to refer to such a corporation as a "global" corporation rather than as, say, a "multi-local" or "multi-national" one? What exactly does it mean when climatologists talk of "global" warming when, amidst an overall planetary warming, some places are actually experiencing cooling? What does it mean to talk about a war on "global" terrorism if acts of violence and "counter-terrorism" occur in quite "local" settings – say, particular streets in London or Baghdad or individual courtrooms in the US or Europe?

Changing theoretical trends within Geography and the broader humanities and social sciences have also had a bearing upon matters of scale. In particular, the reaction by some against Marxism, with its alleged master narrative of global class struggle, and the engagement by many with postmodernism and post-structuralism, with their focus upon difference and local variation, has led numerous writers to place greater analytical attention on the local and to laud its supposed emancipatory potential. Consequently, many on the political left have argued that the local is somehow inherently more democratic than are other scales of decision-making and that a focus upon local difference allows subjects to break free of the oppressiveness of "one size fits all" spatial models of economic and political development which have been part of nationalist modernizing projects and/or Marxist rhetorics about the need for global revolution. Such a view is expressed by the myriad calls to "think globally" yet to "act locally" that have became popular with much of the left, a fascination by many with Foucault's (1986) concept of local "heterotopias," or the arguments of people like bell hooks (1990: 342), who suggested that local "spaces of marginality" are much more than "a site of deprivation [but] also a site of radical possibility, a space of resistance."

Whilst many on the political left were focusing upon the imagined liberatory possibilities of the local in the face of other scales such as the national and global, many on the political right also began to show renewed interest in the local. This was particularly so in the United States, where conservative economic theoreticians argued that responsibility for economic and political decision-making should be devolved to local political authorities – a phenomenon referred to as the "New Federalism." For conservatives, such a devolution of power to the local level was seen as essential to stimulating the market, for it allowed competition between communities in ways which nationally focused political policies and regulations did not. Hence, whereas national wage agreements, especially if backed by the force of law, do not allow workers in different communities to compete against each other on the basis of differences in remuneration rates, having local wage agreements clearly does. Such localization, many conservatives believe, results in markets operating in a manner that more closely approaches a perfect competitive state, since economic actors are no longer fettered by regional or national government regulations.

Concomitant with this problematization of scale has been the explosion of a new vocabulary with which to talk about matters scalar. Thus, the literature during the past two decades or so has been replete with discussion of "the politics of scale," "the production of scale," "the construction of scale," "scalar framing," "scale jumping," "scalar strategies," "scalar restructuring," "scalar fixes," "scale bending," "rescaling" and "descaling," "upscaling" and "downscaling," the "scale division of labor," and what "glurbanization" and the "glocalization" of social processes and phenomena might mean. Given such an explosion of writing on matters of scale, then, in this book I seek to do two things. First, I hope to provide an overview of debates concerning how issues of geographical scale have been thought about within Geography and more broadly, for the spatial resolutions at which social processes take place and are perceived to take place, together with how we talk about them, have significant implications for understanding our world. Hence, the book's primary goal is to serve as a pedagogical resource for students and researchers who wish to "get up to speed" on such theoretical issues. But, second, I hope that, through examining such theoretical debates, the book will also push them in new directions, raising new questions (even if not providing answers!).

In terms of its structure, following a general discussion in Chapter 1 of issues of ontology and discourse as they relate to understanding and thinking about scale, the book contains five substantial chapters, each of which addresses and problematizes a different scale. There is also a brief conclusion. In the text's main corpus I have focused on the scales of "the body," "the urban," "the regional," "the national," and "the global." Certainly, there are others I could have chosen to spotlight, such as the scales of "the household," "the neighborhood," "the continent," and "the province"/"state." However, I concentrate upon the scales explored here because they have been, I think, more subject to theorization by geographers during the twentieth century than most others. This is largely, I would suggest, because they are so central to how life is structured under contemporary capitalism. I should also point out that I have chosen not to devote a separate chapter to "the local," both because this scale is less spatially coherent than others and because I address issues of "the local" when contrasting it with "the global" in Chapter 6. Equally, I do not address scale in its technical, cartographic sense (for a good overview of the latter, see McMaster and Sheppard 2004). Although cartography is clearly important in shaping geographical imaginations concerning how our world is or is not scaled – how cartographers use scales to produce world maps can have signal influences upon how we think various parts of the planet are networked together, for instance – in this book I focus upon how scale is used both as an ordering device with regard to social, cultural, political, and economic life and as a socially produced material entity. My focus, then, is on how various geographical scales come to be, how concepts of scale are deployed as tools with which to think, and how, in the process, they both help us to make sense of the world around us yet can also be used to constrain how we comprehend that world. Finally, I focus principally upon how scale has been conceived within Human Geography and the social sciences more broadly, rather than Physical Geography or the natural sciences. This reflects both my own positionality as a Human Geographer/social scientist and, frankly, the fact that, except for the case of the region, most of the theorizing about scale's status in the twentieth century has occurred in Human Geography and the social sciences.

It is important here also to say a word about the process of constructing a narrative about scale (I return to this issue at the end of Chapter 1 and periodically throughout the book). All texts are narratives. Part of the issue with which any book must deal, then, is the relationship between what is being described and the narrative used to do so. As Sayer (1989: 271) has suggested, the need to structure narrative in a way that makes it comprehensible often means separating objects that are inescapably connected, "even though it risks concealing their interconnections." So it is with what follows. Hence, although they are dealt with separately for purposes of exposition, it is important to recognize that the scales addressed herein are, in fact, deeply interconnected as part of a continuum of social existence and praxis, with each scale shaping others. For example, the processes which create global scales of economic organization are constituted by individual human bodies, whilst such global processes in turn shape how human bodies can develop in very real ways. Likewise, although it is often argued that globally focused organizations such as the World Trade Organization undermine national-level sovereignty, these organizations can only come into being through the acquiescence of various nation-states. However, the necessity of creating a linear narrative – a book's physical structure means that one page must relentlessly follow another – means that a kind of narrative essentialism emerges, in which things that are intimately interconnected and always in a state of flux are discussed as separate entities in successive chapters as if they were discrete and fixed.

Having said this, it is also important to recognize that in the ever-dynamic world of scale-making and remaking, processes of scalar structuration often do crystallize into fairly long-standing "scalar fixes," such that particular scales present themselves as somewhat discrete and enduring entities within a continuous process of historical transformation. This is because, as Lefebvre (1976: 69) has argued, each geographical scale experiences three interconnected moments of existence: first its historical "formation," then its (always partial and provisional) "stabilization," and finally its possible "bursting apart" or transmogrification. In focusing, for purposes of exposition, on the moment where particular scales appear to be stabilized, therefore, it is crucial to not forget the other two moments of their existence – scales are always in a process of becoming and there is always a tension between tendencies towards stabilization and those towards destabilization, even if at any one moment one of these appears to have won out.

In some ways, then, in dedicating specific chapters to specific scales and moving through each in a scalar progression, what follows is a fairly traditional organizational approach to understanding scalar divisions and one critique will no doubt be that it reinforces a sequential view of scales. Equally, though, the book's structure does mirror the scalar progression that has typically been used within the literature – from the body, to the urban, to the regional, to the national, and on to the global – and so, I would argue, is a useful place to start when trying to outline the debates that students might find useful. (Indeed, as I have written the manuscript I have been painfully aware of this tension between, on the one hand, trying to provide clarity of understanding concerning issues of scale by following scalar progressions that will be familiar to students and, on the other, challenging unquestioned conceptions of scale that they may harbor by disregarding the familiar progression and adopting another. For better or worse, I have chosen the former, though my hope is that by highlighting the issue here I have at least raised the question of such a progression in readers' minds and might suggest that, if the book is assigned to classes, different groups of students could be delegated to read the chapters in different orders with the purpose of discussing how each group's understandings of the issues perhaps vary, based upon the order in which they have read the chapters.) Although each chapter focuses on a particular scale, my hope is that the book as a whole will be thought of as somewhat like a movie. Hence, in a movie, each frame, when examined individually, allows us to home in on particular aspects of its construction, aspects which appear fixed in time, even as the totality of the narrative can only be understood by comprehending frames relationally – a sense of dynamism is only possible by seeing each frame in relation to every other one in the movie. Finally, in each of the chapters I have outlined how a particular scale has variously been theorized as a scale, how it has been discursively represented and with what consequences (as I detail in Chapter 2, for instance, the Zoque Indians of southern Mexico do not see the human body as separate from the land in the way in which Western European Enlightenment tradition does, with significant consequences for land-use patterns), and also how it has shaped, and been shaped by, the material world in which we live.

Lastly, in the course of writing this book I have accumulated a number of debts. I would first like to thank Sarah Holloway and Gill Valentine for inviting me to write it. Thanks, too, to the reviewers of the initial proposal and, particularly, of the first draft of the completed manuscript for suggestions that have improved it (even if I have not taken all of them). Also, I thank the staff at Routledge/Taylor and Francis. My constant delays with the manuscript were met with understanding by Andrew Mould, although I fear they may have run off Michael Jones – Michael was my point of contact throughout much of the project, though towards its completion moved on to other things and handed it over to Faye Leerink, who ably brought things to completion. Thanks to Wendy Giminski for draughting the diagrams, and to the Department of Geography at UGA for paying for them and the cost of reproducing copyrighted materials. Thanks also to Jim Wheeler, who read Chapter 4 and gave me good feedback. I dedicate this book to my wife, Jennifer Frum, who was engaged in her own writing project during the course of much of its genesis and of whose accomplishments I am always immensely proud.

Andrew Herod
Athens, Georgia
January 2010

NOTE

1 Throughout, the capitalized term "Geography" and its derivatives (e.g., Regional Geography) refer to the academic discipline.

1

WHAT IS SCALE AND HOW DO WE THINK ABOUT IT?

Main entry: **Scale**
Function: *Noun*
Etymology: Middle English, from Late Latin *scala* ladder, staircase, from Latin *scalae*, plural, stairs, rungs, ladder; akin to Latin *scandere* to climb.

Merriam-Webster Dictionary

The question of scale and of level implies a multiplicity of scales and levels.

Henri Lefevbre (quoted in Brenner 1997: 135)

If the recent explosion of writing on globalization is any indication, matters of scale and the supposed rescaling of social life are on the intellectual agenda in Geography and other disciplines in a big way. Perhaps this is nowhere more evident than in discussions of globalization. Thus, although the myriad interpretations of globalization are quite different in many ways, by and large they share one overriding similarity: they argue that the contemporary economic, political, cultural, and social processes that are taken to be emblematic of globalization are rescaling people's everyday lives and identities across the planet in complex and contradictory ways and are generally undermining other scales of social life, cultural identity, and economic and political sovereignty (see Herod 2009). Hence, much has been made of how globalization is supposedly leading to a weakening of national sovereignty, as national-level govern-ment powers are reassigned to supranational entities such as the European Union and to subnational levels of government. Likewise, many have argued that local communities' power to resist "globally organized" capital's predations has been undercut as globalization has unfolded.

Conversely, whereas many have argued that globalization heralds the evisceration and colonization of scales such as the local, some have suggested that the global's increasing power is bringing with it not the undermining of the local but, perhaps paradoxically, its reassertion. Hence, Barber (2001) contends that globalization and localization are dialectically related processes – the more the forces of cultural homo-genization spread across the planet through the export of, primarily, US popular culture, the more people in various parts of the planet seek to express their differences as a way of maintaining their local cultural identities. Simultaneously, others have explored how cultural practices have supposedly become increasingly hybridized or glocalized, as when a "global corporation" such as McDonald's adapts its menu to take local tastes into consideration.

Such arguments raise interesting questions about the *gestalt* of scale (i.e., how various scales fit together into a coherent whole), includ-ing: what is the relationship between "local" processes and "global" ones? what is the relationship between "local" forms and "global" processes, and vice versa? and, is "the global" simply the sum of all things "local," or is there a degree of synergism, such that "the global" is, in fact, more than the totality of everything "local"? (For an exploration of how such questions play out with regard to language, see Box 1.1.) In light of such questions, in this chapter I explore two sets of issues concerning geographical scale and how it is conceptualized. First, I examine arguments concerning scale's ontological status – that is, debates about whether scales really exist or not. I also outline how the corpus of writing on matters scalar developed historically within Geography. Second, I consider how various scalar metaphors have been used within Geography and beyond, and with what implications. The exploration of these two sets of issues then serves as the basis for the discussion of a number of separate scales in the remainder of the book.

Box 1.1   WHAT MAKES A LANGUAGE "GLOBAL"?

Thinking about how various languages are described raises interesting questions concerning matters of geographical scale. For instance, what does it mean to say that English is a "global language"? Does it mean that it is a very commonly spoken language? If so, then it falls far behind Mandarin Chinese, which is spoken by more people but tends not to be considered "global" in quite the same way. Does it mean that it is spoken in parts of the globe beyond its country of origin? If so, then might we also consider Yoruba and Quechua "global languages," since both are spoken by individuals living across the globe, especially in communities with large numbers of immigrants from Nigeria and Peru? If not, then what makes English a "global language" and these others "local" or "regional" languages? Is it simply that a traveler is more likely to find someone with a passing familiarity with English than with these other languages when traveling? If so, presumably this would vary by their geographic location – a visitor to certain parts of Latin America and the Caribbean (where varieties of Yoruba are spoken amongst diasporic communities) or to a Peruvian immigrant community in, say, Los Angeles may find Yoruba and Quechua more useful than English. Equally, how do the geographically specific and linguistically quite distinct varieties of "national" Englishes – British English, American English, Canadian English, Australian English, New Zealand English, Indian English, Nigerian English, Jamaican English, South African English, Singaporean English, and the myriad Englishes spoken in other parts of the world – meld together to form a "global English"? Is "global English" merely the sum of its constituent "national" parts and, if so, what happens when different varieties come into conflict with one another over word usage, pronunciation, and grammatical syntax?

Furthermore, in what ways is it even possible to speak of "national" Englishes such as "British English" or "American English" when in both countries there are noticeable local variations in lexical practice, intonation, and sentence structure? Thus, if we are to ask how various "national Englishes" come together to form a "global English," then we should probably also ask how various local or regional Englishes come together to form, say, "British English" – in other words, what is the relationship between "local" or "regional" Englishes and

continued

Box 1.1 continued

a "national" English? Moreover, given how certain localities' variants of English dominate what is considered their "national" English, what does this mean for how particular varieties of "local" English from specific regions of any English-speaking country may come to dominate "global" English? In the case of British English, for instance, it is historically the English from the southeast which has dominated the national variant but which has also been the basis for many other "national" Englishes – Australian English, for instance, largely derives from the working-class London English of the eighteenth and nineteenth centuries, given that convicts from this part of Britain numerically dominated those transported to Australia. Rather than thinking, then, that "national" English represents some kind of geographical average of all the local or regional Englishes spoken within a country and that one moves up a kind of linguistic spatial hierarchy from "local" English through a largely geographically homogeneous "national" English to "global" English, in fact much of the English that is taken globally to be representative of, say, "national" British English is from quite specific parts of Britain. Hence, rather than a linguistic hierarchy that links the local, through the national, to the global, it is possible to imagine a hierarchy in which the national is largely bypassed and the English spoken locally in the southeast of Britain has historically dominated much of the English spoken globally.

Equally, we might ponder how particular words from other languages and localities are transmitted spatially and incorporated into a local or national variant of English before becoming differentially rebroadcast globally – the words *shufti* (slang in parts of Britain for "a quick look") and *shampoo* were both introduced into British English by soldiers returning from the Middle East and India, but whereas *shampoo* is commonly understood by most speakers of English worldwide, the fact that *shufti* is unlikely to be known much outside Britain tells us something about the spatiality of word loans and language projection globally. Likewise, the term *boonies* (short for *boondocks*, meaning "in the middle of nowhere") was introduced into American English during the US occupation of the Philippines and was derived from the Tagalog word *bundok* ("a mountain"). As such,

continued

Box 1.1 continued

it is commonly understood by most Americans but would probably be unfamiliar to most English speakers outside either the Philippines or the US. The fact that some borrowed words become part of the "global" language whereas others remain largely national or even local in their usage in the adopting language, then, raises interesting questions about the geography of language and, in this case, its scaling.

Finally, the nationalization of certain local varieties of a language to form a "national standard" can sometimes be organic and can sometimes be planned through the intervention of the national state, a fact that has important scalar implications. Thus, in the case of Italian, the Tuscan dialect was selected by the government after unification to serve as the basis for standard Italian, largely because Florence was a center of the Renaissance and the 1582 founding place of the *Accademia della Crusca*, which had published early Italian dictionaries (see Bertinetto and Loporcaro 2005). In this instance, then, a particular local area's language variant was elevated quite consciously to the national stage. In other instances, national *lingua francas* emerge organically, as an amalgam of various different local/regional languages or dialects that may then subsequently be formally adopted by governments as "national languages."

ON SCALE'S ONTOLOGICAL STATUS

Early interventions

Scale has long been one of Geography's core concepts. Nonetheless, prior to the 1980s – and despite some efforts in the early twentieth century to examine critically the constitution of scales such as "the urban" and "the regional" (see Chapters 3 and 4) – scale was largely a taken-for-granted concept used for imposing organizational order on the world. Whereas both physical and human geographers – as well as many other natural and social scientists – had frequently employed scales such as "the regional" or "the national" as frames for their research projects, looking at particular issues from, say, a "regional scale" or a "national scale," they had generally spent little time theorizing the nature of scale itself.

Rather, researchers typically had simply imagined the world as inherently hierarchically compartmentalized, with scales such as "the regional," "the national," and "the global" conceptualized as natural geographical units/ spatial echelons (from the French *échelon* [rung of a ladder], which itself comes from *échelle*, meaning both "a ladder" and "scale"), as simply the most logical way in which to carve the world up into manageable pieces for the purposes of analysis, or as no more than handy mental contrivances for ordering the world. To all intents and purposes, though, in such approaches scales were seen simply as tools for geographically circumscribing "a relatively closed . . . system, the majority of whose interactions remain within its boundaries" (Johnston 1973: 14) – they were viewed as what Lefebvre (1974/1991: 351) evocatively termed "space envelopes."

Generally, such approaches drew upon the conceptual approach outlined by eighteenth-century German philosopher Immanuel Kant (1781/2007). Whereas Isaac Newton (1687/1999) had viewed time and space as real and absolute things which form a container for natural and social phenomena and processes, Kant argued that neither time nor space were objective, real things but were, instead, subjective constructs through which humans make sense of the world. For Kant, any order appearing in the world is the result not of material processes but of the categorization imposed on it by our brains. Although historically Kantianism has infused much writing in Geography (May 1970; Livingstone and Harrison 1981), it was perhaps Hart (1982: 21–22) who most forcefully articulated a Kantian view of scales when he suggested that they are merely "subjective artistic devices." Given that he viewed them as, essentially, mental fictions, for Hart there could thus be "no universal rules for recognizing, delimiting, and describing" scales, whilst his argument that scales are "shaped to fit the hand of the individual user" encouraged a theoretical stance which viewed the absolute spaces of the Earth's surface as capable of being more or less arbitrarily divided up into bigger or smaller areas, with little concern for how such areas might relate to anything "on the ground." Whilst there are many examples of works adopting such conceptual formulations, Peter Haggett's *Geography: A Modern Synthesis*, arguably one of the most influential texts of the 1970s' "spatial science" tradition within Geography, epitomized this approach to scale. Thus, Haggett (1972/1975: 17) used a scalar schema for dividing up the world that relied principally upon a fairly arbitrary mathematical progression through what he called "Orders of Magnitude"

– his Fifth Order of Magnitude represented areas on the Earth's surface between 1.25 km and 12.5 km in diameter, his Fourth Order areas between 12.5 km and 125 km in diameter, his Third Order areas between 125 km and 1,250 km, his Second Order areas between 1,250 km and 12,500 km, and his First Order anything with diameters from 12,500 km to 40,000 km, the planet's approximate equatorial circumference. For Haggett, the important analytical questions were not how scales are delineated or made but how "changes in scale change the important, relevant variables" (Meentemeyer 1989: 165) as they affect various processes and phenomena, whilst the key theoretical declarations involved arguing that multiscalar analysis is crucial for understanding the complexities of human and natural systems. However, following the publication of two articles by Taylor (1981, 1982) and of Smith's (1984/ 1990) book *Uneven Development*, the concept of scale began to be hotly debated within Human Geography – and, to a degree, Physical Geography – and continues to be so today.

Drawing upon world-systems analysis, Taylor (1981) argued that particular scales take on certain roles under capitalism. Specifically, he maintained that: the global scale is the "scale of reality," the scale at which capitalism is organized; the national scale is the "scale of ideology," as it is the scale at which the capitalist class primarily promulgates class-dividing ideologies (such as nationalism); and the urban scale is the "scale of experience," for cities are where everyday life is primarily lived in capitalist societies. Taking this further, he subsequently argued that there were fundamental contradictions with regard to the scales at which socio-economic classes have historically organized (Taylor 1987). Hence, under capitalism, classes "in themselves" have often been defined globally, such that it is possible to talk analytically of a global working class and a global capitalist class. Classes "for themselves," however, have tended to organize nationally, regionally, or locally. For Taylor, then, there was a disconnect between the scales at which classes under late industrial capitalism exist and the scales at which they often perceive themselves to exist. Despite offering important analytical insights, though, Taylor's approach suffered from two lacunae: i) by suggesting that certain scales played particular roles within how capitalism operates, his argument seemed to present a somewhat functionalist approach to scale; ii) it did not have much to say about how scales come about in the first place, for it focused instead upon how they are used once in existence (for more on world-systems analysis, see Box 1.2).

Box 1.2 WORLD-SYSTEMS ANALYSIS

World-systems analysis is an approach to understanding how the world is structured economically which draws upon neo-Marxist theory. The approach is probably most associated with the work of sociologist Immanuel Wallerstein (1974), who argued that with the emergence of capitalism there developed a core and a periphery to the world economy and that the former has grown at the latter's expense – during the nineteenth century, for instance, industrialization in Europe was fuelled by colonialism, with colonies in places such as Africa both providing raw materials for European factories and serving as markets for their products (Herod 2009). Consequently, Wallerstein argued, the world economy can only really be understood by looking at it from a global perspective, a position which privileges the global scale of analysis. A number of other writers have been closely associated with this approach, including Samir Amin, Giovanni Arrighi and Andre Gunder Frank (see Amin *et al.* 1982). However, these writers have often disagreed on the specifics of the world system's emergence. Hence, whereas Wallerstein dated its emergence to the so-called "long sixteenth century" (approximately 1450 to 1650) when capitalism began to emerge in Western Europe, Frank argued that its origins were in the pre-capitalist era and that it had already substantially emerged by the fourth millennium BCE, by which time humans had populated all of the continents except Antarctica.

If, for Taylor, the key questions were what roles various scales play under capitalism, for Smith they concerned how the various scales at which capitalism is organized came into existence. In seeking to understand why capitalism's economic landscape was decidedly unevenly developed, Smith suggested there is a fundamental geographical tension at play within the very structure of capital itself which leads to the production of various geographical scales. Drawing upon Marxist theory, Smith argued that capitalists must constantly negotiate the tensions between two opposing needs: the need to be fixed in place so that accumulation can occur, and the need to be always on the lookout for more profitable locations for investment and accumulation elsewhere. Thus, on the one hand capital seeks to embed itself in particular locations so that accumulation may quite literally "take place" – capitalists must develop relationships with local suppliers and/or labor forces if they are to produce commodities. Simultaneously, though, capital seeks to level economic space through equalizing the rate of profit across the landscape. This tension between the needs for fixity and for mobility, between efforts to differentiate the landscape through investing in particular favored locations and yet concurrently to equalize the rate of profit across it, Smith contended, is the basis for the "production of scale." The production of particular geographical scales is the mechanism by which capital negotiates its own contradictions.

Having outlined such a theoretical framework, Smith (1984/1990) initially explored three principal scales at which these spatial negotiations play out under capitalism – the urban, the national, and the global – but later added "the regional" to this triumvirate (Smith and Dennis 1987; Smith 1988). (Interestingly, these are the same three scales that Taylor (1981: 6) suggested "dominate most of our thinking in the social sciences.") With regard to the urban scale, Smith argued that there is a certain spatial coherence to urban labor markets, one manifested through a locality's Travel-To-Work Area – the distance that workers will generally commute from their homes to their places of paid work on a daily basis. Smith suggested that daily commuting patterns thus provide a material basis for defining the urban scale, for beyond a particular TTWA the geographical coherence of daily labor markets begins to break up. Smith contended that the regional scale was also based upon labor markets. However, it was not how far workers would generally travel on their daily commute that served as the basis for this particular scale but, rather, how territorial divisions of labor emerged out of the various sectors of the economy. For instance, what has provided the industrial regions of the north of England or South Wales or of north-central Appalachia a certain territorial coherence has been the development of coal mining and steel making. Significantly, such a conceptualization of regions as entities shaped by the sectoral division of labor meant that "the region's" place within the typical hierarchy of scales was not quite as fixed as often previously imagined. Hence, instead of regions sitting between "the urban" and "the national" scales, as they had frequently been conceived to do, it was now possible to imagine them as sometimes also transgressing "the national" scale, as when particular industries were so functionally integrated on two sides of a national border that they operated as a single entity – as with the industrial region of the northeastern US and Ontario, Canada.

In the case of the national scale, Smith suggested that this scale developed out of the need of various capitalists (often initially self-identifying themselves on the basis of particular cultural identities, such as speaking a common language) to cooperate to defend their collective interests against other capitalists who likewise self-identified themselves on the basis of (different) cultural commonalities – British capitalists, for instance, have come together to defend what they perceive as their collective interests against French capitalists. Whilst he recognized that nation-states clearly existed prior to the emergence of industrial capitalism, Smith argued that the rise of capitalism transformed their nature – although scales such as the national scale of social organization were "historically given before the transition to capitalism . . . in extent and substance they [we]re transformed utterly at the hands of capital" (Smith 1984/1990: 135). Hence, he proclaimed, the national scale increasingly came to be a scale at which economies were organized politically, such that groups of capitalists might not only compete with each other at a sub-national scale but also cooperate with one another at a national scale, perhaps to ensure that particular pieces of legislation were passed that limited working-class agitation or kept other countries' capitalists at bay through tariffs. The result was that, by the twentieth century, many had come to see the world economy as something made up of a series of interlocking national economies, with nation-states serving as the spatial containers within which economies were constituted (we shall return to this representation – and its problems – in Chapter 5).

Finally, Smith maintained that the global scale results from capitalists' desire to universalize the wage labor form of production, spreading it to the far reaches of the planet and sweeping aside other means – for example, feudal obligation – by which surplus labor is extracted from one class by another. Whereas, therefore, the planet's physical limits are geologically given, the global scale of capitalist economic organization has had to be actively constructed. Given that capitalist wage forms emerged in particular places – often seen to be early modern Europe – capitalists had to work to make them global, to spread them geographically from their point(s) of origin to the wider world. Hence, transnational corporations (TNCs) do not suddenly arrive on the global stage fully formed but actively make themselves "global" through their investment strategies. Moreover, the nature of their global presence has changed over time. Thus, although initially it was trade and the mechanism of exchange which bound together the planet's separate economies into a world-system (Herod 2009), over time the world economy's spatial integration has increasingly been brought about through manufacturing under conditions of wage labor, as TNCs have stretched their assembly lines across national boundaries (for more on these authors, see Box 1.3).

At about the same time that geographers Taylor and Smith were pondering the nature and significance of spatial scale, sociologist Anthony Giddens was doing likewise, if not in quite such explicit terms. Specifically, in his theory of structuration, Giddens (1984: 119) explored what he called the "regionalization" of space, with regionalization taken to be "the zoning of time-space in relation to routinized social practices." For Giddens, space's differentiation – its regionalization – was central to structuring social life. Regionalization, he argued, was the process whereby "locales" – what he viewed as the "*settings of interaction*" (p. 118, original emphasis) – were internally differentiated, often into "front regions" and "back regions" wherein different types of social behavior take place. Giddens saw such a scalar differentiation of space, which he also linked to matters of core and periphery in the making of the world economy's unevenly developed geography, as resulting from everyday social practices and thus open to constant renegotiation. There were, however, several limitations in Giddens's approach. For example, although he contemplated how social practices are choreographed in time and space, he failed really to theorize the *process* of scale *production* or to appreciate that such production might be a source of social conflict. Consequently, for him scales remained largely descriptive devices for differentiating between various parts of the landscape. Additionally, his use of scalar terminology ("regionalization") was somewhat spatially imprecise, with "regions" seen to range in physical extent from rooms in houses (p. 119) to whole urban areas, to globally defined core and periphery zones (p. 130) – an imprecision which limited its theoretical value. On the other hand, the concept that geographical scales might be produced and reproduced as a result of people's everyday "routinized social practices," although not explicitly explored by Giddens, nevertheless would subsequently be an important element in the ongoing theorization of scale and its nature.

These early writings by Taylor, Smith, and Giddens represented an effort to suggest that there is a material basis to scales – that is to say, scales are socially produced and have real consequences for social life and

Box 1.3 BRIEF BIOGRAPHIES OF SEVERAL SCALE THEORISTS IN GEOGRAPHY

John Fraser Hart, Professor of Geography at the University of Minnesota, was born in 1924 and grew up in Virginia. His work has principally been on the agricultural geography of the South and the Midwest of the United States. He has described Regional Geography, with its detailed descriptions of local landscapes, as being at the core of Geography as a discipline. He is a past president of the Association of American Geographers.

Peter Haggett was educated at Cambridge University and is Emeritus Professor at the School of Geographical Sciences, University of Bristol. Born in 1933, he is best known for being a leader in the quantitative revolution which reshaped Geography, beginning in the 1950s. He has worked extensively on issues of the spatial diffusion of diseases, especially measles and, latterly, AIDS. Arguably, his most famous publication is *Geography: A Modern Synthesis*, first published in 1972.

Peter Taylor is Professor of Geography and Director of the Globalization and World Cities (GaWC) Research Network at the University of Loughborough. Originally from London, he was for many years a professor at Newcastle University in northern England. Trained within the spatial science tradition in Geography, in the 1970s Taylor adopted Marxism as an analytical framework. Probably best known as a political geographer, more recently he has written about global cities.

Neil Smith is Distinguished Professor of Anthropology and Geography, and Director of the Center for Place, Culture and Politics, at the City University of New York. Originally from Scotland, he has spent most of his professional life in the US. He has written extensively on issues of gentrification and urban politics. His first book, *Uneven Development* (1984), explored the production of nature and space under capitalism from a Marxist perspective. In this book he first articulated a theory of the production of geographical scale which served as the basis for much theorizing about scale during the next two decades.

are neither natural nor simply the most logical way in which to partition the world for purposes of analysis, nor are they merely mental conveniences. Significantly, though, whilst writers such as Taylor and Smith were arguing for a materialist understanding of scale, the 1980s also saw the growth of a language of spatial metaphor in the humanities and social sciences – a growth associated with the influence of postmodernism and, especially, feminism (Shands 1999) – in which themes of "subject positionality," "mapping," "location," "place," "marginal spaces," "sites of identity," "crossing boundaries," "giving space" for alternate voices, "grounding," and myriad others became popular in discussing both identities and potential political strategies of liberation (Price-Chalita 1994; Silber 1995; Moore 1997). However, although the growth of a language of space and scale in many academic fields played an important role in shaping developments in social theory, the users of such metaphors generally did not think too seriously about the production of material spaces and scales and had little recognition that the spaces and scales to which they referred might have material referents over which various social actors struggled. The result of such a lack of consideration was that both space and scale were essentially excluded from critical theoretical scrutiny because they were assumed to be unproblematic stages and/or containers of social life. This assumption, and the conceptual detachment between the metaphorical and the material, led Smith and Katz (1993: 75) trenchantly to suggest that "[s]patial metaphors are problematic in so far as they presume that space is not." Hence, the term "crossing boundaries" may not only be a metaphorical turn of phrase to describe the transgressing of categorical borders associated with, say, national identity, but it might also have a material reality to it – as when undocumented migrants seek to cross the boundaries into certain spaces (such as the US space-economy) yet are prohibited from doing so by the US Border Patrol.

To summarize, what crystallized out of these early articles was a growing debate – largely between materialists and idealists – about spatial scales' ontological status. Derived from their rival philosophical traditions (Kantian idealism versus Marxist materialism), the major line of division between these two groups concerned whether scales are real. For idealists such as Hart, scales are not real things that are visible in the material landscape but are simply mental contrivances for circumscribing and ordering processes and practices. They may have metaphorical or categorical purchase, but have little material heft. For materialists, on the other hand, scales really exist as substantive social products. Although in practice those drawing on a materialist understanding of scale may encounter problems in trying to delimit an urban area's or a region's exact boundaries and may, in practice, end up drawing those boundaries in much the same way as does an idealist when delimiting their area of research study, it is important to see this as a methodological – rather than an ontological – matter, for despite their, perhaps, similarity of empirical practice, conceptually these two approaches view scales and entities such as regions in fundamentally different ways.

It is important to recognize, however, that although this broad idealist/materialist division represents a fundamental faultline with regard to conceptions of scale, there has also been something of a division within the materialist camp. Specifically, whereas all materialists saw scales as real things rather than just mental constructs, Taylor's approach tended to conceive of them as part of a pre-existing, perhaps even natural, way in which the spatiality of capitalism is structured – he was primarily concerned with their role within capitalism rather than with their origin. For others such as Smith, though, the focus was not so much upon scales' function – although that was important – as it was upon their creation: in Smith's view, scales do not just exist, waiting to be utilized, but must instead actively be brought into being. Consequently, there is a politics to their construction. Hence, scales must be created and are subject to conflict in their making.

Despite such differences, it is important to recognize that there were also similarities between these camps. In particular, both idealists and materialists generally saw scales as areal in nature, with the scalar boundaries delimiting various spaces serving as containers of particular parts of the landscape – regions, nation-states, etc. As a result, they both tended to view scales as separate and distinguishable entities within a hierarchy of spatial divisions, such that one particular process or social practice could be considered, say, "urban" whereas another could be regarded as "national" in scope. Finally, both idealist- and materialist-inspired writers shared commonalities in that both groups generally conceptualized scalar hierarchies either as having a verticality to them, such that one climbed up from "the urban" to reach other scales such as "the global" (as if individual scales were like rungs on a ladder), or as having a horizontality to them, such that one moved out from "the urban" to other scales (as if individual scales were like concentric circles) (see Figures 1.1 and 1.2).

Figure 1.1 Scale as a ladder

Figure 1.2 Scale as concentric circles

Deepening and widening the debate

Following initial exploration in the 1980s, there has been a veritable explosion of scalar theorizing in the subsequent years. Much of this has come from authors who have drawn explicitly upon Marxist theory or, more generally, political economy and who have approached matters of scale from a materialist point of view. Whereas much initial theorizing had revolved around considerations of how, as Smith (1984/1990: xv, emphasis added) put it, "*capital* . . . produces the real spatial scales that give uneven development its coherence," a second wave began both to solidify arguments concerning the social production of geographical scales and to refine them. In particular, in response to criticism that much theorizing had been capital-centric and seemed to suggest that it would be possible to "read off" various scales' production simply by understanding the logic of capital dynamics (Herod 1991), there was some significant consideration given to how a more nuanced understanding of scale-making might be developed. Although not all of those involved adopted their terminology, much of this reassessment focused upon what Delaney and Leitner (1997) called "the *construction* of scale," in contrast to "the *production* of scale," with the term "construction" being used largely to refer to "bottom-up" and "non-capital-centric" approaches to scale-making, in contradistinction to approaches which saw scales as being produced out of the internal machinations of capital itself.

An early assessment of the debate's state came in Herod (1991), which explored three limitations to some of the early Marxist interventions. First, materialist considerations of scale-making had tended to focus upon capital's dynamics and to downplay other social mechanisms – such as patriarchy – which shape how landscapes are produced. Thus, given that women's daily commuting patterns are typically shorter than men's because of their domestic responsibilities (they have less time to travel to places of paid work) (Madden 1981; Hanson and Pratt 1995), an examination of TTWAs delimited according to typical male commuting patterns and those delimited according to typical female patterns would reveal significant differences in the spatial extent of particular urban areas. Certainly, in making this point, it is important to distinguish between the geographical size of particular urban areas and how we might conceptualize the urban scale itself. In other words, just because a particular city's TTWA expands spatially, this should not be confused with a change in the urban scale's nature. At the same time, though, given that "the geographical limits to daily labour markets express the limits to spatial integration at the urban scale" (Smith 1984/1990: 137), the gendering of labor markets, such that the distances traveled by men and women to locations of paid work is often quite different, it is important to consider if the spatial extent of daily labor markets is to be considered a basis for theorizing "the urban scale." Likewise, the global expansion of the capitalist wage–labor relation has not been gender neutral, as women form the bulk of the laborforce of TNCs expanding into export-processing zones in the Global South.

Second, social agency was rather limited in these early visions of scale-making – scales were seen to emerge out of the internal contradictions of capital rather than as a result of political struggles between, say, capital and labor, between different segments of capital and/or labor, or between other social groups, such as nationalist independence fighters and colonial powers. Third, it was argued that the scale-making process needed to be considered in a much more dialectical manner, with a greater focus not just on how scales are made but also upon how various scales are inter-connected – for instance, a social actor's control over one scale may allow them to shape scale-making at other geographical resolutions, as when the national Conservative government of Margaret Thatcher in the United Kingdom abolished an entire sub-layer of government (metropolitan councils) which were often controlled by its political opponents.

Such arguments were explored in a number of empirical arenas, though particularly in relation to the growth of "Labor Geography" (Herod 2001). For instance, Herod (1997a) examined how conflicts between two rival dockers' unions in the US east coast maritime cargo handling industry in the 1950s revolved around their efforts to construct different scales of representation – whereas one, which was strong really only in the coast's dominant port (New York), argued for a coastwise bargaining unit through which it could use its large numbers in New York to dominate the other ports, its rival, which had support in ports up and down the coast, argued for a system in which New York remained isolated from the other ports. Once these two unions merged, the result-ant solitary union then sought to develop a system of national contract bargaining to replace the port-by-port bargaining that had traditionally dominated the industry, thereby to prevent employers from playing dockers in different ports against each other. Of particular significance in this case was not only the union's ultimate securing of a national con-tract but also the fact that it forced a concomitant nationalization of the employers' bargaining structure, against the will of many (Herod 1997b). Switching focus to the global scale, Herod (1997c) challenged much of the received wisdom concerning globalization, which had tended to assume that it was corporations which were responsible for the integration of the global economy, and showed how various labor organizations had been intimately involved in encouraging this integration. Looking at the rescaling of labor relations from the employer side, Crump and Merrett (1998) showed how US agricultural implements manufacturers replaced national collective bargaining agreements in the 1980s with local-level concessionary contracts as they restructured their organization.

Whereas these works tended to focus upon the realm of industrial employment to show how workers and employers have made new scales of economic and political organization, several authors examined other social actors' scale-making. Miller (1997), for instance, analyzed how peace campaigners in Massachusetts built new organizational scales, linking local communities across the state in an expanding hierarchy of spatial scales. In a similar vein, Brown (1995) outlined how AIDS activists in Vancouver, British Columbia, had constructed new scales of organization. McCarthy (2005), meanwhile, examined how environmental non-governmental organizations went about making new scales of organization and developed a series of "scalar strategies" in light of passage of the North American Free Trade Agreement. Taking the debate in a slightly different direction, Marston (2000) argued that much of the literature had focused on what she called the sphere of production, rather neglecting the spheres of consumption and reproduction. As a corrective, she outlined how late nineteenth/early twentieth century middle-class US women adopted and promulgated new conceptualizations of the home and household, which had implications for how structures and practices of consumption and social reproduction developed at scales far beyond the home (see also Marston 2002).

Meanwhile, in his exploration of contemporary transformations of the world economy, especially as they relate to scalar restructurings of the state, Swyngedouw (1996, 1997a, 1997b) suggested that the two scales which are frequently taken to be at the opposite ends of the scalar spectrum – the local and the global – are so deeply intertwined that, rather than viewing them as separate domains as in the traditional scalar hierarchy, understanding present-day capitalism requires the development of new concepts and language. Hence, championing the portmanteau word "glocal" to express the mutual constitution of the local and the global,

Swyngedouw (1997b: 137) argued that, rather than thinking of processes of planetary restructuring as instances of globalization, it is better to think of them as a process of glocalization, wherein "local actions shape global . . . flows, while global processes, in turn, affect local actions," recognizing all the while that "[o]ther spatial scales are also deeply implicated in these events as well" (p. 138). Beginning with the affirmation that social life is process-based and in a state of constant change, transformation, and reconfiguration, Swyngedouw (1997b: 140–41) declared that starting any analysis "from a given geographical scale seems . . . deeply antagonistic to apprehending the world in a dynamic, process-based manner." Stressing process over form, he argued that "theoretical and political priority . . . never resides in a particular geographical scale, but rather in the process through which particular scales become (re)constituted." Consequently, he reasoned, "scale (at whatever level) is not and can never be the starting point for sociospatial theory" (p. 141). Instead, the starting point must be theorizing and understanding "process," specifically the processes whereby scales are transformed and transgressed through social struggle. Accordingly, he (p. 142) advocated the "abolition of the 'global' and the 'local' as conceptual tools" and suggested, in their place, that analysts concentrate "on the politics of scale and their metaphorical and material production and transformation," declaring that sociospatial struggles produce "a nested set of related and interpenetrating spatial scales that define the arena of struggle, where conflict is mediated and regulated and compromises settled" (p. 160). Perhaps revealing a degree of causal hierarchy in his own thinking, Swyngedouw asserted that in such struggles "[p]lace matters, but scale decides" (p. 144) – thus, for him (1997a: 169), "[i]f the capacity to appropriate place is predicated upon controlling space, then the scale over which command lines extend will strongly influence the capacity to appropriate place."

For his part, Smith also moved away from a capital-centric view of scale-making. Thus, he examined how anti-gentrification activists in New York City's Lower Eastside made linkages with activists in other neighborhoods, thereby uniting various neighborhood groups into a broader citywide movement (Smith 1989a, 1993) – through what he called "scale jumping," activists were able to create a new scale of urban politics. Likewise, in his discussion of European statecraft he traced how various political actors have been central in the making of nation-states and, perhaps, a post-national Europe in the form of the European Union (though see Chapter 5 for more on the complex relationship between individual nation-states and the EU) (Smith 1995). Finally, he expanded the spectrum of scales within his schema to include "the body," "the home," and "the community" (Smith 1993). With regard to the first of these scales, Smith particularly assessed the ability of homeless people to cross urban space more readily using a "Homeless Vehicle," a device (largely made out of a supermarket shopping trolley) by which they could move their belongings with them and even sleep in, so keeping out of the elements – the vehicle enabled homeless people to overcome their confinement to fairly circumscribed areas within particular neighborhoods caused by their inability to travel far from their meager belongings or, perhaps, a shelter. He further underscored how the body was also typically seen as the site or scale at which individual identity is expressed, at least in Western cultures. Drawing on this argument, Smith suggested that "the home" was the scale at which routine acts of social reproduction (eating, resting, child-rearing) generally take place and in which individual bodily identities are often formed. Lastly, he averred that "the community" was the scale at which collective identities are often formed.

Several non-geographers also engaged in debates over geographical scale and how its making has been conceptualized. For example, Macrae (2006), a political scientist, examined how the scales at which gender-related policies within Europe are promulgated have been transformed in recent decades – whereas such polices were once "firmly grounded in the politics of the nation-state," they have now been "'rescaled' and partially relocated to an EU level" (p. 522). Mills (2006), also a political scientist, investigated maternal health policies in Mexico to show how the rescaling of health services dedicated to maternal health – specifically, a decentralization designed to bring healthcare "closer to the people" – has had a series of unintended consequences that, in some cases, have actually made it more difficult for women to access such services. For her part, Masson (2006), a sociologist, analyzed how women's movements in Québec, Canada, have reworked the scales of their own collective action in response to provincial government projects to "downscale"/ devolve government functions through creating new regional policy-making and policy-management institutions. The result has been that whereas the various regions within Québec were once little more than territorial subdivisions of the province, the sub-provincial region has now "been materially and discursively constructed by Québec women's movement actors as a legitimate and relevant scale for feminist politics"

(p. 462). In similar fashion, Guenther (2006) showed how the women's movement in Rostock, eastern Germany, refocused its scalar attentions away from the national state, which was seen to be largely unfavorably disposed toward feminist demands, and turned its attention to organizing at a citywide and statewide level.

Putting all of this together, the consideration of scale and its making that came in response to the early writings of people such as Taylor and Smith represented a deepening and a widening of the debate, as various commentators sought both to include other actors beyond capitalists in the process whereby scales are produced and to move beyond the sphere of production. Nevertheless, these "deepeners and wideners" generally shared several commonalities concerning how they thought about scales, commonalities both with each other and with some of the writers against whom they were writing. First, all saw scale as socially constituted, even if some preferred a language of "social production" whereas others preferred one of "social construction." Second, they largely conceived of scales in areal terms, with scales understood as geographical boundaries around particular spaces. Third, scales were regarded as fluid – that is to say, although particular scales may be made and fixed at particular times and places, it should not be supposed that they are unchanging or unchangeable. Finally, they frequently focused upon the relationship between different scales, either in terms of how social actors "jump scale" as they move from working at one spatial resolution to another, or in terms of how various social actors seek to control one scale so as to dominate processes, actors, or phenomena at another.

Questioning scale (again)

If deepening and widening has been one aspect of scalar discussion since the 1990s, a second – one that generally has taken place in the 2000s – has been a more fundamental reconsideration of what scales may or may not be, how the relationships between different scales are conceptualized, and the language surrounding the "politics of scale." This discussion has been made up of several more or less connected elements. Some of the more minor of these have involved what we might consider tying up some loose ends of the extant debate. For instance, Chapura (2009) suggested that too much focus had been placed on intentional scale-making efforts and questioned how the scales which develop out of the unintended consequences of social action might also be theorized.

Likewise, Herod (1997b) argued that the language of "scale jumping" had unintended consequences because it suggested that social actors were bounding between pre-made scales that were just waiting to be used, a representation that appeared to contradict the contention that scales are actively created through social praxis. Finally, there was concern that whilst scales such as the national or the global were viewed as having to be actively brought into being, others (such as the scale of the body or the local) seemed to be considered less socially produced scales, almost as if they were a natural or default basis from which to build other scales (for instance, Brown and Purcell (2005) show how political ecologists have often privileged the local scale, which is itself a problematic scale of analysis because of its lack of spatial precision (I discuss this in Chapter 6)). Thus, whereas a discourse of "becoming" permeated much of the analysis concerning creation of scales such as the global – as with examinations of how corporations or labor unions have "become global" – this was less true about some other scales and led to suggestions that these latter were being privileged analytically as the foundational scales upon which all others are constructed. In opposition to this privileging, some commentators argued that scales such as the body and the local are no less socially produced than are any others – hence, manufacturers must work to embed themselves in communities by establishing linkages with local suppliers and workforces, thereby actively becoming local (Cox and Mair 1988), and human bodies are at least as much socially produced as they are biologically.

Beginning in the late 1990s, though, there was also a more deep-seated reconsideration of some of the arguments about scale and how different scales could be conceptualized in terms of their relationships with one another. One attempt to think about scale in a slightly different way was Cox's (1998: 2) distinction between what he called "spaces of dependence" ("those more-or-less localized social relations upon which we depend for the realization of essential interests [, for] which there are no substitutes elsewhere [, and which] define place-specific conditions for our material well being and our sense of significance") and "spaces of engagement" (which he took to be those spaces in which, and through which, social actors build linkages with actors located elsewhere). Drawing such a distinction provided the basis for Cox's argument that scales materialize out of the social relations that connect one particular actor's local spaces of dependence with those of others situated at some distance from them, all as part of a strategy of engagement with them. What is perhaps most significant about Cox's approach is that he put forward a networked – rather than areal – vision of scale. Certainly, Jones (1998) and Judd (1998) took him to task by querying what he meant by terms such as "localized social relations," since "localized" merely seemed to beg the question of scalar precision. Nevertheless, Cox's approach suggested a view which saw moving from one scale to another not as "a movement from one discrete arena to another" (Cox 1998: 20) but as a process whereby social actors develop networks of associations which enable them to shift amongst and between different spaces of engagement. Rather than conceiving of scales as relatively discrete areal entities (even if constantly remade through everyday actions à la Giddens), Cox maintained that they should be conceived of in networked terms.

Such ideas of scales as networked have been further explored by others. Thus, Latham (2002) highlighted the transnational movements of various entrepreneurs in the New Zealand restaurant industry to question whether they have "gone global" through their travels or whether, in bringing overseas skills and culinary traditions to Auckland, they have actually "localized" themselves in New Zealand – or even done something different altogether. Rather than a scalar language focused on geographical areas and spatial echelons (what Castree et al. 2007 refer to as a "topographical" view of scale; see also Amin 2002, 2004), Latham (p. 116) drew on actor-network theory to make a case for a "topological" view of scale, one which sees scales not as areal units but as parts of networks in which "places are simultaneously made as both local and global, without necessarily being wholly either" (Box 1.4 distinguishes topographical from topological approaches). Significantly, in so doing he argued that this approach did "not . . . claim that the world lacks structure or hierarchy." Instead, Latham wished to distinguish between approaches which consider scale as central to how space and time are ordered – in this regard he pointed specifically to Smith's (1993) claims that scale is a central dynamic in the generation of "difference" – and those which deem size and scale "relational effects," "uncertain effect[s] generated by a network and its modes of interaction" (Thrift 2002: 40). Suggesting (p. 138) that in "seeing scale as one of the central elements of spatial differentiation . . . scale theorists, in fact, blunt our sense of how spatial difference is produced and maintained," Latham intended "not to throw away the concept of scale [but] simply to be more sceptical of its importance and analytical purchase" (p. 139).

Box 1.4 TOPOGRAPHICAL VERSUS TOPOLOGICAL APPROACHES TO SCALE

Topographical approaches to scale view each scale as the boundary line that encloses a particular absolute space – that of the region or the nation-state, for instance. The boundaries enclosing such spaces form an unbroken perimeter around each one, in much the same way that contour lines on a map do not cross each other and are unbroken. Topological approaches, on the other hand, do not imply enclosure of spaces but instead describe how networks are structured, usually in terms of lines and nodes. In such approaches, scales are represented not by the enclosure of different-sized absolute spaces but by the relative length of the lines connecting various nodes – longer lines are generally taken to represent the global, whilst shorter lines represent the national or regional scales.

Likewise, Conway (2008: 207), in her analysis of how the World March of Women (WMW – "a newly emergent and innovative transnational feminist network") has attempted to make connections between women's organizations in different places and organized at different scales, expressed concern "with the spatial connotations of the transnational and the strange geographical vacancy that has accompanied increased usage of the term in feminist contexts" and suggested that many feminist theorizations of transnationalism had been "hampered by the absence of a vocabulary of scale or of spatiality more generally." Through her analysis she examined "the geographic grounding and specificity of varying claims to the transnational [, particularly] the relation of the transnational to the so-called 'local.'" Specifically, Conway (p. 224) averred that "the transnational" – which she distinguished from "the global" – should not be thought of as a single, fixed scale, in terms of either its geographic reach or its scope, but that it could "imply many geographic scales (sizes, levels, relations), both within and across the juridically recognized borders of nation states" and that "[a]s with any representation of scale, any instantiation of the transnational is constantly being produced through sociospatial practices and is therefore somewhat fluid." Thus, the WMW could be conceived of as simultaneously local and global, whilst the idea of fluidity and the image of scales as networked and as "more horizontal and less hierarchical and [as] characterized by greater reciprocity, dialogue, mutual respect, and recognition" (p. 223), she argued, "invokes an alternative socio-spatial imaginary of both the global and the movement as rooted in places/locales that are dispersed, diverse, and increasingly densely networked in a huge variety of ways, rather than as single and unitary." Significantly, Conway identified the "recognition of the multiple geographic sites and scales of struggles, the irreducibility of their existence and their significance, and the displacement of a hierarchy of scales of movement practice" as central elements in "creating a postcolonial politics."

A second line of retheorizing came out of Jones's (1998) critique of Cox's networked approach. Specifically, although she agreed "that we must not take scale for granted" (p. 28), she also suggested that it should not be merely assumed that "scale exists as an ontological category" (p. 27) and "that we may be best served by approaching scale not as an ontological structure which 'exists,' but as an epistemological one – a way of knowing or apprehending" (p. 28). Whereas Cox and others had argued that scales are constructed as real things – in other words, they have ontological weight – and that, for example, a case of actors shifting their focus from, say, neighborhood- to metropolitan-level scales of action represents a remaking of material scales of organization, Jones (p. 27) maintained that "we must also accept that scale itself is a representational trope, a way of framing political-spatiality that in turn has material effects." Continuing, she reasoned that "if scale is a trope, then we can no longer see it as neutral or transparent in how it represents [, for e]very trope carries with it its own rhetoric, its own ability to shape the meaning of space." Thus, using the example of how urban planners have promulgated practices such as land-use zoning, Jones contended that individuals have increasingly come to know and understand urban environments not through their direct experiences of what is immediately around them but through envisioning, as if from some "God's-eye" viewpoint, each neighborhood as being part of a larger metropolitan entity. The result has been that the "more that urban information [has been] presented through maps and zones, the more the city [has been] understood only by way of these sorts of spatialized and geometrical systems, until what was considered 'true' about the city was altered in practice." Consequently, she (p. 28) insisted, the "creation of scale as a trope for understanding the city did not merely shift politics from one level to another [– from the immediate neighborhood to the city as a whole – but] recast what was true or knowable about the city within the frame of scale[, such that c]ertain questions about the city simply became un-askable." Such a transformation did not represent "simply a jump in scales [but was] a fundamental change [in how] the city was known and apprehended." This recognition is important, for as a representational trope "scale may be implicated in enabling particular relationships of power and space that advantage some social groups but disadvantage others" (p. 28). In response, Cox (1998: 44), whilst recognizing that scale is indeed "a representational practice," countered that, like space, "scale is constituted by, but is not reducible to, objects" and that, as a result, scale "makes a difference through the way it conditions, as spatial arrangement, the realization of the causal properties of the objects that make it up." Thus, he maintained, TNCs may frame their actions in particular scalar terms, but they still must build material space-crossing organizational networks that link places near or far in concrete ways. For Cox, scale may serve as an epistemological structure but it is also ontologically real.

If questions concerning whether scales are ontological realities or merely epistemological ones were one set of issues raised, others related to how the language of scale had itself been used. Drawing inspiration from French Marxist Henri Lefebvre (1974/1991), sociologist Neil Brenner (2001: 592) argued that much writing on the topic of space and scale had used the term "politics of scale" (a phrase coined by Smith, in the Afterword to the 1990 edition of *Uneven Development*) rather imprecisely and that there had been a "noticeable slippage in the literature between notions of geographical scale and other core geographical concepts, such as place, locality, territory and space." Although he directed his comments at the literature generally, he took as his point of departure Marston's (2000) paper, in which she had suggested that "the household" was an important yet overlooked analytical scale, as it represented a scale shaped principally by processes of reproduction and consumption – processes largely ignored in the discussion up till that time. In particular, Brenner suggested that the utility of the concept of the "politics of scale" was being diluted through its use to refer to things that could just as easily be considered to be the "politics of *place*" or the "politics of *territory*." In its place, he sought to distinguish between what he saw as two different usages of the term "politics of scale" in the extant literature.

Specifically, Brenner (p. 599) contended that many used "politics of scale" to refer to conflicts concerning "the production, reconfiguration or contestation of some aspect of sociospatial organization *within* a relatively bounded geographical arena." In such a formulation, the term "scale" was essentially interchangeable with "boundary," whereby one unit – say, the urban – could be distinguished from another – say, the regional. Scales were simply relatively differentiated and self-enclosed geographical containers. By way of contrast, Brenner (p. 600) declared that in its second meaning "scale" had been used to refer to "the production, reconfiguration or contestation of particular differentiations, orderings and hierarchies *among* geographical scales." Here, the term "politics of scale" referred to "not only the production of differentiated spatial units as such, but also, more generally, [to] their embeddedness and positionalities in relation to a multitude of smaller or larger spatial units within a multitiered, hierarchically configured geographical scaffold." In this second sense, the referent was not the scalar unit itself and the political actions within it but "the *process of scaling* through which multiple spatial units are established, differentiated, hierarchized and, under certain conditions, rejigged, reorganized and recalibrated in relation to one another," with scale understood here primarily as a "modality of *hierarchization* and *rehierarchization* through which processes of sociospatial differentiation unfold both materially and discursively." Drawing upon this distinction, Brenner (p. 604) averred that the term "politics of scale" should be employed in a more judicious way to refer specifically to the creation of "a hierarchically ordered system of provisionally bounded 'space envelopes' . . . that are in turn situated within a broader, polymorphic and multifaceted geographical field," whilst the venerable geographical terminology of "locality," "place," "territoriality," and the like could continue being employed to speak about the practices whereby sociospatial organization *within* relatively bounded spatial units is produced. In an effort to develop a terminology through which to clarify his distinction, Brenner professed that the first meaning of the term "politics of scale" might better be described through what Jonas (1994) has called "the scale politics of spatiality" whereas "the politics of scaling" might best express its second.

Having made this distinction, Brenner (pp. 604–8) outlined eleven maxims by which he sought to clarify the scale debate, these being:

1   Scalar structuration is a dimension of sociospatial processes – i.e.,
    the structuring of scales is best understood as an outcome of

sociospatial processes (e.g., capitalist production, state regulation) than as an inherent property of a society's spatiality.

2 Processes of scalar structuration are constituted and continually reworked through everyday social routines and struggles – scales are never made "once and for all" but are constantly remade through everyday practices.

3 Processes of scalar structuration are dialectically intertwined with other forms of sociospatial structuration – the scaling of social relations represents only a single aspect of their spatiality, such that, for instance, nation-states are not only scaled (generally having national- and local-level tiers) but they also have territorial expression and varying degrees of power centralization, with all of the implications for core-periphery relationships that this entails.

4 There are multiple forms and patterns of scalar structuration – different sociospatial processes (such as the accumulation of capital or the regulation of the nation-state) are likely to have quite different forms of scalar structuration.

5 Scales evolve relationally within tangled hierarchies and dispersed interscalar networks – the meaning and function of any particular scale can only truly be understood relationally, in terms of its vertical and horizontal relationships to other scales.

6 There are multiple spatialities of scale – although scales have often been thought of in areal terms, such territoriality represents only one dimension of their sociospatiality and ignores others forms, such as networks.

7 Scalar hierarchies constitute mosaics not pyramids – processes of scalar structuration do not produce a single nested scalar hierarchy but are better understood as a "mosaic of unevenly superimposed and densely interlayered scalar geometries" (p. 606), such that the meaning of the global or the urban will vary historically and geographically.

8 Processes of scalar structuration generate contextually specific causal effects – depending upon the specific context within which sociospatial processes unfold, a given process may generate different types of scalar effects at different times and in different places.

9 Processes of scalar structuration may crystallize into fairly long-standing "scalar fixes" – despite the assertions in maxim 2, certain aspects of daily life may become enframed within relatively stable geographical hierarchies (such as those within the organs of the

nation-state) in which social practices organized at certain scales predominate.

10 Established scalar fixes may constrain the subsequent evolution of scalar configurations – once certain scalar fixes are established they may then shape how other scales develop which, in turn, may influence how these first scalar fixes further evolve.

11 Processes of scalar structuration constitute geographies and choreo-geographies of social power – scales may operate both as the arenas of power and social struggle, and as the objects of such.

In response to this implicit critique of their work, Marston and Smith (2001) replied jointly and suggested that Brenner's effort to distinguish conceptually the "production of space" from the "production of scale" made a certain degree of sense and that the two should not be conflated – as they put it (p. 615), "scale is a produced societal metric that differentiates space; it is not space *per se*." Nevertheless, they argued, it was impossible to separate entirely these two concepts, for the "production of scale is integral to the production of space" (p. 616). They also contended that, paradoxically, in his critique Brenner had done the very thing of which Marston had been critical, namely failing to take seriously matters of social reproduction and consumption. Brenner and Smith, at least, also seemed to have quite different views with regard to the nature of scales – Brenner's (2001: 605–6) assertion that "scales evolve relationally within tangled hierarchies and dispersed interscalar networks," such that "scalar hierarchies constitute mosaics not pyramids," was in singular contrast to Smith's (1992a: 73) claim that "social life operates in and constructs some sort of nested hierarchical space rather than a mosaic" and that, "[w]ith a concept of scale as produced, it is possible to avoid on the one hand the relativism that treats spatial differentiation as a mosaic, and on the other to avoid a reified and uncritical division of scales that repeats a fetishism of space."

Furthermore, in seeking to distinguish between a "politics of scale" that refers to how political struggles play out within various relatively differentiated and self-enclosed geographical units and a "politics of scaling" that refers to how such geographical units are structured together in some kind of scalar configuration, Brenner seemed to imply that it was possible to draw a fairly sharp distinction between the processes occurring within areal units and those linking them together. Put another way, for such a conceptual proposition to be sustainable, the scalar boundaries around such spatial units must be seen as sufficiently impermeable as to stop processes internal to these units from bleeding across the boundaries that circumscribe them and playing a role in interscalar configuration. However, ultimately, this view may say more about how such boundaries are conceived of and represented discursively than it does about their ontological form – for instance, as Mitchell (1998: 90) has argued, during the twentieth century nation-states were largely pictured as separate and discrete entities, "with each state marking the boundary of a distinct economy," although in practice their borders were fairly fluid and "national" economies were, in fact, quite penetrated by overseas capital (Herod 2009). For his part, Purcell (2003) suggested that both parties in the debate were simply talking past each other.

Adopting a related, if distinct, approach to thinking about the relationship between scale and spatiality, Collinge (2005: 190) explored how Marxist writers had conceptualized scales as emerging out of a dialectical relationship between space and society but argued that this approach was problematic because the concept of a sociospatial dialectic (Soja 1980) itself was problematic. Claiming (p. 201) that such a concept presents a faulty dualism in which space and society are held separate conceptually so that they might subsequently be brought together in a dialectical interaction, Collinge averred (p. 194) that the "distinction between society and space, whether conceived dichotomously or dialectically, has generally operated as a hierarchical duality." Indeed, he continued, these approaches' interpretive frameworks cannot work without first discursively representing society and space as separate, if related, objects. The problem, Collinge (p. 201) maintained, is that when scale's social production for the purposes of trying to control and/or make space in certain ways is stressed, one side of the dialectic is privileged, resulting in "a hierarchical distinction in which 'society' is the causally dominant term and 'space' is its essential-but-malleable other." On the other hand, when scale's spatial configuration for the purpose of shaping social relations and bounding social objects is the focus, the other side of the dialectic is stressed, leading to the assertion of "an inverse hierarchy in which space is dominant and society is its erased other." The result, Collinge sustained, is a situation in which scale becomes a "Janus-faced third term that stands between society and space."

Delving into the realm of semiotics, Collinge argued (pp. 201–2) that, in its Janus-faced capacity looking simultaneously toward both the spatial and the social, scale serves concurrently as "a (substantive) boundary line that attempts to secure [society and space] as separate spheres and to mediate their dialectical interaction, and [as] a (conceptual) conduit that subverts this separation and mediates their semantic confusion." Scale is thus, for him, what Jacques Derrida calls a "between" term, one which both blurs the boundary between the social and the spatial through promulgating the notion that they are dialectically interpenetrated and "stands between the opposites 'at once'" (Collinge, quoting Derrida 1981: 212) by implying that there is nevertheless some distinction between them – the social production of space and space's recursive shaping of social processes and relations, in other words, are perceived as two distinct moments within a unified whole. Given that in Marxist approaches scales are conceptualized as emerging out of the sociospatial dialectic, for Collinge (p. 202) scale in such approaches "becomes an undecidable term, neither unambiguously social nor spatial nor for that matter consistently sociospatial." Consequently, he maintained, rather than drawing upon the Marxian Lefebvrian tradition, writers on matters scalar would be better served by drawing on a post-structuralist and non-dialectical Derridean approach in which scale "gives society and space the form each . . . needs to have 'presence'" (p. 202). Put slightly differently, Collinge declares that rather than resulting from the dialectical interaction of society and space, scale is actually an ordering mechanism through which each of these realms establishes its own separate discursive identity. For Collinge, therefore, "[s]caling is . . . what Derrida has called an infrastructural term that (like 'spacing' and 'writing') draws attention to the coincidence of bounding and unbounding processes." Scale should therefore be thought of, he argued, in Platonic terms as a productive chōra (Box 1.5), a semiotic "receptacle of becoming" which provides the necessary pre-ordering of the social and the spatial that precedes their own signification – that is, their recognition as separate realms.

There are two important issues concerning Collinge's critique of Marxist understanding and his call for a post-structural approach to scale based in semiotics. First, it is questionable whether analysts writing in the Marxist tradition actually do see society and space as separate ontological entities or whether they simply refer to the two separately because of limitations of language. Hence, Smith (1981: 113) had earlier maintained that approaches which see space and society as separate entities within a dualism are not Marxist in origin but, rather, draw upon positivist systems of categorization based upon Kantian idealism (they present what he calls a "dialectics in words not action"), whilst

---

Box 1.5 PLATO'S CHŌRA

Plato described the *chōra* as a place that sits between the world of ideas and the material world. It is a place in which ideas are materialized, a "receptacle of becoming." However, because the *chōra* occupies a zone between the realm of ideas and of materiality, it is neither. Consequently, although it is understood in light of the idea/material dualism through which Plato believed the Universe to be constituted, it simultaneously undermines that binary. In terms of considering scale as *chōra*, Collinge is arguing that scale allows both the social and the spatial not only to be seen as separate realms but also to be understood as deeply imbricated within each other – in other words, the social/spatial binary is concurrently reaffirmed and undermined.

Castells (1983: 311) has argued that space is not separate from, nor merely "a 'reflection of society,' [but that] it is society." This raises questions about the relationship between the material and the ideational and issues of narrative, namely that, whilst the social and the spatial may be understood to be not separate realms ontologically, language limitations can result in them being written about as such. It is crucial to remember, then, that discursively presenting two things as separate for purposes of elucidation is quite different from accepting that they are ontologically separate. Second, conceptualizing scale as an "infrastructural term" suggests that scales are not really material entities in the landscape but are, instead, simply discursive constructions for categorizing society and space.

Expurgating scale?

If Brenner's efforts to deconstruct the term "politics of scale" represented one reevaluation of the literature, and Collinge's call for adoption of a Derridean view of scale represented another, the provocative intervention by Marston et al. (2005), in which they argued for "expurgat[ing] scale from the geographic vocabulary" (p. 422) and abandoning notions of scale altogether, almost seemed to bring the debate full-circle (see also Jones et al. 2007; Marston et al. 2007). Thus, they averred, extant views of scales as forming a vertical hierarchy were problematic because they generally reinforced a privileged position for certain scales within the hierarchy – typically the global (usually considered to be "at the top" of the hierarchy) or the local (often considered to be the scale upon which all others are built). Unhappy with such a privileging, Marston et al. suggested they had three intellectual positions by which they could respond: i) they could reaffirm such a hierarchy but argue that greater attention be paid to those other scales which sit between the hierarchy's two ends; ii) they could develop some hybrid models in which scalar verticality is supplemented with a degree of scalar horizontality (through, for instance, combining it with some form of network theorizing); or iii) they could abandon the concept of hierarchical scale in its entirety. Arguing that pursuing the first option would reinforce an ontological–epistemological nexus in which one particular spatial dimension (verticality) is privileged, and developing some kind of hybrid would simply add a second nexus in which a different spatial dimension (horizontality) is also privileged, they proposed instead a flat (as opposed to horizontal) ontology, for this, they maintained, "does not rely on any transcendent predetermination – whether the local-to-global continuum in vertical thought or the origin-to-edge imaginary in horizontal thought" (p. 422). Whilst they accepted that scales may serve as part of an epistemological ordering frame – calling something "national" or "global" can radically influence how it is conceptualized – they asserted that this is quite a separate matter from claiming that the landscape itself is materially organized into various nested spatial hierarchies. Such a "flat ontology," they reasoned, allowed them to "discard the centring essentialism that infuses not only the up-down vertical imaginary but also the radiating (out from here) spatiality of horizontality" (p. 422) and to adopt a conceptual position in which various parts of the Earth's surface (what they called "sites") are viewed as interlinked but are not seen to be in any kind of spatially hierarchical – either vertical or horizontal – relationship. In such an approach, scales may have material effects but they are, ultimately, simply part of a representational trope and do not exist in any material sense.

As might be imagined, Marston et al.'s paper elicited a wide range of responses, from claims that their formulation simply mirrors Actor-Network Theory's propositions (Collinge 2006), to those declaring that they had misinterpreted the distinctions between ontology and epistemology (Hoefle 2006), to those who insisted that their conceptualization was little more than a repackaged Kantianism. Furthermore, given their parting comments – "if [in adopting a flat ontology] we lose the beauty of the 'whole thing' when we downcast our eyes to the 'dirt and rocks,' at least we have the place – the only place – where social things happen, things that are contingent, fragmented and changeable" (p. 427) – it did seem that, even as they were arguing against privileging particular scales, they were actually privileging the local ("the only place where social things happen" presumably being the place right at one's feet, to which one's downcast eyes turn). Indeed, Jonas (2006: 402) pointed out that, paradoxically, short of a world in which no concepts or points of view are "privileged" and all are equally valued, Marston et al.'s position could be interpreted as simply privileging different concepts from those hitherto allegedly privileged – or, as he put it: "[b]y replacing scalar constructs with a site-based epistemology, Marston et al. seem to be privileging non-scalar representations and categories over and above spatial (scalar) concepts and identities."

Taking a slightly different tack, Kaiser and Nikiforova (2008: 537–38) made a case that "writing scale out of human geography will [simply] help to hide the social constructedness of scales and the way they are discursively deployed to naturalize and sediment a set of sociospatial relationships through everyday practices." In fact, they predicted that,

> exercising scale as a subject of critical inquiry . . . will almost certainly contribute to the stabilization of both scale ontologies and the hierarchical power relations partially based on them, in that scales will more easily return to the naturalized, taken-for-granted categories of analysis that they were perceived to be in the past . . . the very thing that Marston et al wish to eliminate from academic and political discourses and practices.

Ironically, then, Marston et al.'s approach could be read as a deeply conservative one. For their part, Kaiser and Nikiforova wanted to keep a sense of scale as something that has real material effect but suggested that the most productive way to do so was through exploring scale's "performativity." Drawing upon Judith Butler's (1990) work, they proposed a post-structural alternative to the approaches which they felt had so far dominated scalar debates, an alternative which "retain[ed] scale as a critical subject of sociospatial inquiry in human geography" yet resolved some of the shortcomings they believed Marston et al. had identified. This approach would focus upon constructing a "political genealogy of scale ontologies," by which they meant "an historically contextualized study of the naturalizing and sedimenting production of scaled knowledge, in order to expose the power relationships lying behind the truth claims about scales and scalar hierarchies" (p. 538). The key is to understand scales not "as things in the world and as the actors that matter" (p. 540) but as "a category of practice" and to focus attention on "the enacted discourses through which scales become . . . [and] materialize through the repetition of sets of citational practices that stabilize as well as challenge the boundary, fixity, and surface effects that materialize" (pp. 541–42).

Likewise Moore (2008: 213), whilst agreeing that "scale is an epistemological, rather than ontological, reality," also suggested that Marston et al. were throwing the baby out with the bathwater. For him, Marston et al.'s proposal reproduced a binary way of thinking that suggested that "denying the ontological reality of scales implies that they are merely inconsequential heuristics in the minds of geographers that 'do no work,' or have no effect in themselves." This is especially so, he argued, because by maintaining that social scientists have only three options available to them – maintaining the hierarchical view of scale, with minor adaptations; developing some kind of hybrid vertical/hierarchical scalar models; or discarding scale as a concept in favor of a flat ontology – Marston et al. reveal their position to be one in which they are focused upon scale's theoretical status rather than upon how scalar practices are engaged in in everyday life. In contrast, Moore argued that "it is not necessary to retain a commitment to the *existence* of scale in order to analyse the *politics* of scale." For Moore, the key problem with extant views of scale is the failure "to make a clear distinction between scale as a category of *practice* and [as a] category of *analysis*" (p. 203). In response, he outlined six broad issues which arise if scale is treated as a practical category rather than as an ontological entity and which contribute to the development of what he called a "non-substantialist approach to scale."

First, even if viewing scale as an ontological reality is abandoned, it can still be claimed to be an epistemological one. Drawing an analogy with the concept of "the nation," Moore suggested (p. 214) that "scalar narratives, classifications and cognitive schemas" can constrain and/or enable particular "ways of seeing, thinking and acting." Thus, much as "the nation" is an ontological fiction which relies for its coherence upon the idea of an imagined community (Benedict 1983), the idea of its existence as an ontological reality shapes the actions of millions of people on a daily basis – a fact revealed when they engage in nationalist rhetoric or when, say, "British people" are accused of acting in an "un-British way." For Moore, the key issue is not to ponder whether scales have ontological existence but, rather, to consider how various social processes and phenomena are viewed as scaled and what this means for conceptualizing the world and for how people behave in response to having internalized such scales. Second, Moore argued that even if scales have no ontological hierarchy – whether vertical or horizontal – à la Marston et al., the fact that people believe that they do has real consequences for how individuals behave. Hence, nation-states spend much time, money, and effort convincing their citizens that there is indeed a hierarchy of local, regional, and national government, each with different responsibilities and powers. Furthermore, depending upon the particular nation-state form, they will often place the individual citizen's body (another scale) rhetorically at the center of the nation-state and suggest either that all other levels of the state should be responsive to it – as in Lincoln's famous 1863 Gettysburg proclamation concerning "government of the people, by the people, for the people" – or that the individual citizen's body should be submissive to the power of the state, as in fascist regimes.

Third, even if scales are conceptualized as not having ontological existence, this does not mean that the fact that they can be conceived of in different ways plays no role in shaping how people interact with the world. Hence, the metaphors which are chosen to represent how the world is scaled can have significant implications for people's behavior – for instance, is the world presented as scaled like a ladder, with one climbing up from the local to other scales, or is it presented as scaled like a series of concentric circles, with one moving out from the local to other scales? (we shall return to such matters in the section below.) In this case, although neither of these representations may be "more accurate" in some ontological sense, they can have quite varied implications for how we think of the world and our relationship to other people, places, and things. Fourth, scales may be used proactively as part of a broader spatial politics. Hence, many TNCs seek to present themselves as "global" when confronting workers (in the belief that workers will consider it more difficult to challenge them), yet as "local" when trying to convince their consumers that they are not a foreign firm. Thus, how scales are used to frame issues is important. Fifth, scale should be thought of not so much as a category of analysis but as one of practice, such that scales are variable and constantly in flux, rather than simply being "fixed" and "refixed" (as when the state is seen to be restructured by a shift in emphasis from the national to the local scale, with both conceived of as pre-existent entities between which power moves). Finally, he insisted that considerations of scale need to place primacy on processes and relations over substances, entities, or things, so as to avoid reifying scales.

As one of the few non-geographers to be much engaged in discussions of scale, Jessop (2009: 89–90) has argued that the debate to date has resulted in several "scalar traps," which have come from the failure to distinguish between three separate "scalar turns": the *thematic* turn (when scholars, rather than exhibiting "scalar indifference," come to view scale as a key category in analysis); the *methodological* turn (when scale is conceptualized as a useful entry point for analysis); and the *ontological* turn (when scale is recognized as a key element in the structuring of the natural and social worlds). Jessop identified three consequences associated with this failure. First, there is what he calls "scalar conflationism," in which writers "fail to distinguish among (a) scale as a relational property of social relations, (b) phenomena *conditioned* by scale in this sense, by its causal processes, and by its emergent effects on non-scalar aspects of the real world; and (c) non-scalar factors *relevant* or implicated in the production of scale" (p. 89). The result is that scale has emerged as something of a "chaotic conception" (Sayer 1984), "found anywhere and everywhere." Second, there has been "scalar reductionism," in which scales are used to explain non-scalar features of a particular process or phenomenon, an approach in which other dimensions of spatiality – place, territory, or network – are largely ignored and the key question is simply to identify the relative importance of different scales. Third, there has been "scalar essentialism," in which scale "is abstracted from its associated substantive content, with the result that the aggregate, cumulative causal powers of a given object are attributed to its scalar properties alone." This essentialism, Jessop suggested, typically occurs in one of three ways: i) ideationally, as when scale is viewed as an a priori mental category or metric à la Kantian-inspired approaches; ii) materially, as when scale is automatically and/or implicitly regarded as the primary aspect of all social relations; and iii) fetishistically, as when scale is considered to exist independently of the social or natural relations and processes which create it and the causal powers of any natural or social object are attributed to its position in any scalar hierarchy.

To avoid these scalar traps (and to avoid rejecting scale as a useful category), Jessop averred, several steps are required. Clearly, the most obvious is simply to recognize them. But this is not sufficient. Consequently, Jessop indicated that it is important to investigate scale as a socially produced dimension of spatiality, and to recognize that scales are produced out of many different processes which may be complementary or may be contradictory. Exploring how these different scaling, rescaling, and descaling processes may come together yet may also work at cross purposes is useful, Jessop (p. 91) sustained, because: it "avoids the view that scale is an external material constraint and/or an a priori mental category; it interprets scale as an emergent constraint that results from social action and recognizes the variability of spatial horizons of action; and it allows for scalar selectivity and scalar-selective activities." In particular, he suggested that what is needed is to develop a conceptual vocabulary that is sufficiently rich as to examine, across a range of spatial dimensions (such as place, territory, and network), spatial imaginaries, and horizons of action, how such different moments and elements of spatiality can be explored from different entry points – thematic, methodological, and ontological. Through adopting what he called a "spiral movement," as first one and then another moment and element of spatiality (place, territory, network, etc.) is stressed from each of these three conceptual portals, Jessop (pp. 91–92) sought to articulate an approach that "would enable investigators to explore the social world from different entry points while still ending with an equally complex-concrete analysis of the current conjuncture in which each aspect of spatiality finds its appropriate descriptive-cum-explanatory weight" – in other words, in which no aspect is privileged a priori beyond its actual explanatory significance. Finally, he argued (p. 93) that the third step is "to relate scale to structure, process, imaginaries, and agency in a comprehensive critical realist, strategic-relational approach," one which recognizes "not only the multidimensional nature of spatiality but also the necessary and inherently complex articulation of spatiality with temporality." However, in seeking to develop an approach in which no aspect of spatiality is privileged beyond its actual explanatory significance, Jessop both assumed that it would, in fact, be possible to determine such explanatory significance and downplayed the fact that, in combination, the various aspects of spatiality he identified might produce a synergy greater than the sum of each individual aspect.

Perhaps in response to some of these developments, both Brenner and Jessop, as self-professed "previous advocates of a scalar turn" (Jessop *et al.* 2008: 389), have more recently advocated an approach which questions the privileging of any single dimension of sociospatial processes, scalar or otherwise, and which promotes in its stead "a more systematic recognition of polymorphy – the organization of sociospatial relations in multiple forms – within sociospatial theory." Suggesting that the past three or four decades have witnessed the development of four distinct spatial lexicons – territory, place, scale, and network – in the social sciences, each closely intertwined theoretically and empirically and associated with specific spatial turns, Jessop *et al.* have maintained (p. 391) that there has been relatively little exploration of "the mutually constitutive relations among those categories and their respective empirical objects." Rather, advocates of a given turn have often privileged ontologically a single dimension, "presenting it as the essential feature of a (current or historical) sociospatial landscape" and have viewed other forms of sociospatial organization simply as "presuppositions, arenas, and products of social action." As a consequence, "attempts to establish the primacy of a given sociospatial dimension [have] tend[ed] to expand its analytical and empirical scope to encompass an ever broadening range of phenomena," such that the "carefully defined abstractions of territory, place, scale, and network are . . . rendered increasingly imprecise, and may even be transformed into chaotic conceptions." Although they examined such essentialism in all four spatial arenas, for our purposes here what is particularly important is that they argued that even if networks – which are typically viewed as having a fairly flat morphology – may have become more important materially in the past few decades, this is still not sufficient to justify the adoption of a "flat ontology" à la Marston *et al.* Thus, although what they called "network-centrism" – in which "a one-sided focus on horizontal, rhizomatic, topological, and transversal interconnections of networks, frictionless spaces of flows, and accelerating mobilities" prevails – has frequently dominated analyses of various social processes, adopting such a representation is different from saying that the world is materially constituted as flat.

As a way to avoid privileging a single aspect of spatiality, Jessop *et al.* developed what they called "the TPSN [Territory/Place/Scale/Networks] framework," which they used as a heuristic device for exploring these four elements' "differential weighting and articulation in a given spatio-temporal context" (p. 393). In doing so, they contended that each sociospatial concept can be deployed in three ways (see Table 1.1). Thus, using the example of "territory," they declared that territory can be explored:

- in itself (territory → territory), as a product of bordering strategies – that is, as a geographical domain created and bounded by struggles over space between and amongst competing interests;
- as a structuring principle or causal mechanism (territory → place; territory → scale; territory → network) whose form can shape that of the other elements of sociospatiality;
- as a structured field (place → territory; scale → territory; and network → territory) produced, in part, through the impact of other sociospatial structuring principles on territorial dynamics.

Although they did not themselves do so, following Jessop et al.'s lead we can perhaps elucidate how such a matrix might play out with regard to scale. Hence, scale might be explored:

- in itself (scale → scale) – that is, as a spatial entity created through struggles between and amongst competing interests as they engage in their various scaling strategies;
- as a structuring principle (scale → territory; scale → place; scale → network) that molds how other aspects of sociospatial relations develop – thus, scales shape how territories are structured (as in the relationships between different levels of government), how places are constituted (as in how various locales are located within particular spatial and scalar divisions of labor as a result of decisions made elsewhere (perhaps at a global headquarters situated a continent away)), and how networks function (as in how power flows from the core to the periphery of any particular network);
- as a structured field (territory → scale; place → scale; and network → scale) as territories, places, and networks influence how scalar dynamics unfold – thus how scales are made and operate is shaped by, for instance, the relationships between different elements of government (local v. national), how different places are located within a broader spatial and scalar division of labor (what, for instance, does this mean for workers' ability to make common cause across space?), and how flat or hierarchical are different networks (Waterman (1993), for instance, has argued that whereas labor solidarity traditionally

Table 1.1 A matrix of interaction between territory, place, scale, and networks as structuring principles and as fields of operation

Structuring principles	Fields of operation			
	Territory	Place	Scale	Networks
Territory	Past, present, and emergent frontiers, borders, and boundaries	Distinct places in a given territory	Multilevel government	Interstate system, state alliances, multi-area government
Place	Core-periphery, borderlands, empires, neomedievalism	Locales, milieux, cities, sites, regions, localities, globalities	Division of labor linked to differently scaled places	Local/urban governance, partnerships
Scale	Scalar division of political power (unitary state, federal state, etc.)	Scale as area rather than level (local through to global), spatial division of labor (Russian doll)	Vertical ontology based on nested or tangled hierarchies	Parallel power networks, non-governmental international regimes
Networks	Origin-edge, ripple effects (radiation), stretching and folding, crossborder region, interstate system	Global city networks, polynucleated cities, intermeshed sites	Flat ontology with multiple, ascalar entry points	Networks of networks, spaces of flows, rhizome

Source: Jessop et al. (2008: 395)

involved a very hierarchical set of relationships, as workers in one country wishing to make contact with those overseas generally had to contact their national union, which then contacted an international union organization, which then contacted a national union organization in the second country, which finally put them into contact with overseas workers, the proliferation of computers and the increasing ability of workers to have one-on-one contact through resources such as email has resulted in an organizationally flatter model of labor transnationalism)?

Drawing upon this schema, Jessop *et al.* submitted that the various interactions in each of the framework's cells should be understood "as expressions of diverse attempts at strategic coordination and structural coupling within specific spatiotemporal contexts" (p. 396). Maintaining that their TPSN framework illustrates the limits to extant scalar theorizing, they argued that their approach demonstrates how the relative significance of territory, place, scale, and networks as structuring principles for sociospatial relations can vary significantly historically and contextually. This means that different strategies of, say, capitalist crisis-resolution will undoubtedly entail a range of attempts to reorder the relative importance of these four sociospatial dimensions and their associated institutional expressions as new modes of capital regulation are explored. Equally, the significance of each of these dimensions is likely to change as new organizational structures and strategies emerge. Hence, Jessop *et al.* contended, much of the work on, for instance, the political economy of capitalist regulation "could be reinterpreted as showing how territory, place, scale, and networks were sutured in historically and geographically specific configurations to forge the Fordist-Keynesian spatiotemporal fix, and that, after a period of trial-and-error searching, experimentation, and contestation, new TPSN combinations seem to be emerging that are more suited to a postnational, unevenly developing global economy" (p. 397).

Taking a similar line of argument, Leitner *et al.* (2008: 158) suggested that "there has been a tendency [in Geography and the broader social sciences] to privilege a particular spatiality – only to abandon that [subsequently] in favour of another" as fashions in sociospatial theorizing change. Thus, with regard to analyses of social movements, they pointed out that whereas in the early 1990s the literature "became replete with such ideas as scale-jumping, scalar and multi-scalar strategies," by the late 1990s/early 2000s, "with recognition of the limits to scalar thinking and calls to abandon scale, scalar tropes ha[d] been replaced by networks" (although it should be pointed out that an interest in networks has a much longer history within Geography – Haggett and Chorley, for instance, published their book *Network Analysis in Geography* in 1969). More recently still, an increasingly favored spatial trope has been that of mobility as part of what Sheller and Urry (2006) have called "the new mobilities paradigm," a paradigm which has supposedly developed in response to the "mobility turn" (Adey 2010) within the social sciences. Although they focused upon a slightly different set of spatialities (place, scale, networks, mobility, and sociospatial positionality) from Jessop *et al.*, Leitner *et al.* nevertheless beat a similar drum, declaring (p. 158) that "*a priori* decisions (ontological or otherwise) to reduce this multi-valency to any single master concept can only impoverish analysis, by offering a partial viewpoint into how geography matters" in political struggles and that "it is necessary to pay attention not only to the pertinence of particular spatialities in particular contexts, but also to their co-implication." Consequently, it is not just a matter of these spatialities being co-present in any particular situation which is important but "also how they shape one another" and, thereby, social movements' trajectories.

After illustrating how each of these five spatialities has been variously explored by theorists, Leitner *et al.* used the example of the Immigrant Workers' Freedom Ride (IWFR) – a September 2003 event in which immigrant workers and activists set out for Washington, DC, from ten cities across the US, stopping in 103 cities and towns before reaching the capital – to show how "multiple spatialities are co-implicated and co-constitutive in complex ways during social movement struggles, with unpredictable consequences" (p. 166). Specifically, they demonstrated how the IWFR's activities can be read from the perspective of each of these different spatialities and how those involved engaged with each of these different kinds of spatialities as part of their spatial praxis – mobility (as the activists traveled across the country), place (as they stopped in various communities), networks (as they linked up with local networks engaged in supporting immigrants' and workers' rights), scale (as they built coalitions with local chapters of various national organizations), and sociospatial positionality (as they transformed themselves into state-less human beings by disposing of all identification documents such as driver's licenses when challenged by various agents of the territorially bounded nation-state). For Leitner *et al.*, then, a variety of spatialities (place, scale, networks, positionality, and mobility) have resonance for social movements' geographical imaginaries and material practices. Although geographical scale is one of these, their "complexities and . . . co-implication" (p. 169) mean that "the spatialities of contentious politics cannot and should not be reduced to scale or any other spatial 'master concept.' No single spatiality should be privileged since they are co-implicated in complex ways, often with unexpected consequences for contentious politics."

The intersection of scaled territorial (i.e., areal) units with scaled networks has also been explored by Swyngedouw (2007), in the context of analyzing political ecological questions, specifically the "production of nature" (Smith 1984/1990). Arguing that the social production of nature is an integral aspect of producing scales, Swyngedouw explored the politics surrounding efforts by the Franco dictatorship in Spain to develop an irrigation and hydro-electric power scheme that was national in scope, transferring water from some parts of the country to others as part of a grand modernization program. In particular, he showed how the transformation of the Spanish landscape through the large-scale movement of water across it was predicated upon specific, yet contradictory, scalar reworkings. Hence, Franco sought to encourage national integration through dam building and irrigation projects. However, doing so required a rescaling of his vision from the national to the supranational, as Spain became increasingly integrated into various US-dominated international organizations in the post-1945 era. Through this work, Swyngedouw argued (p. 11) that "scale is not ontologically given, but [is] socio-environmentally mobilized through socio-spatial power struggles." Thus, explanatory priority "never resides in a particular social or ecological geographical scale [but] in the socioecological process through which particular social and environmental scales become constituted and subsequently reconstituted" (Swyngedouw and Heynen 2003: 912; see also Swyngedouw 2004).

The past decade or so, then, has seen a large number of writers explore scale's nature, in the process advocating various different epistemological approaches – some drawn from Marxism, some from feminism or postcolonial theory, and some from other theoretical approaches. However, it is also important to recognize that, regardless of how scale is conceived of ontologically, the ways in which the relationship between various scales is presented metaphorically and visually can influence in profound ways how we conceptualize the world around us and its social and natural processes. It is to such matters that I now turn.

METAPHORS OF SCALE

Metaphors shape how we think about the world in important ways. Typically, they allow us to take from one context an image with which we are familiar and apply it to another context to try to make sense of things with which we are less familiar or things which are complex and which we are seeking to make clearer. For example, the Victorians often used the metaphor of the steam engine to describe the human body's operation – muscles were its pistons, food its fuel, lungs its boilers, and so forth. Whilst the principal reason for so doing was that such a metaphor made it easier to understand a complex entity (the body), the choice of the steam engine was largely because that was a technology with which they were familiar. Interestingly, as technologies have changed, the metaphors used to describe the body have also changed. Thus, today the body's operation is often illustrated using computing metaphors – the brain, for instance, is seen as the "central processing unit." Indeed, Mulgan (1991: 19–20) has argued that the proliferation of computer technology has done much to shape ways of thinking across a wide spectrum of knowledge:

> The spread of electronic networks has been matched by a widespread use of the network as a logical device or metaphor, something that is good to think with ... Computers have done much to spread familiarity with the idea of logical rather than physical space, with their topological representations of flow diagrams, branching trees and other patterns ... [W]hereas in the eighteenth and nineteenth centuries the workings of the brain or of societies were conceived as analogous to those of the loom or the steam engine, both are today conceived as complex networked systems for producing and processing information.

It is important to bear in mind, though, that a change in metaphor usage does not necessarily imply a material change in what is being described. Bodies work in the same way, regardless of which of these two metaphors we choose as descriptive aids. Choosing one metaphor over another, therefore, is usually not done on the basis of which is an empirically more accurate representation of a process or phenomenon but, instead, is based on how someone is seeking to understand that process or phenomenon and/or the context within which such choices are made – living in the steam versus the computer age.

The importance of different scalar representations, then, is not that they are necessarily reflective of some underlying material reality but, rather, that they provide an entry point for engaging with the material world. In this context, several metaphors have been popular descriptors of how the world is structured. Arguably, the already touched on ladder metaphor and the scales as circles metaphor have been the two most commonly used metaphors for describing scalar hierarchies. Thus, in the ladder metaphor (and here we should not forget that the English word "scale" comes from the Latin *scala*, meaning "a ladder"), each scale is seen as a particular rung, such that there is a clear distinction between, say, "the urban" and "the national," and there is a well-defined verticality implied in the relationship between them – one quite literally climbs up and down the scalar hierarchy. Hence, certain scales are considered above some and below others – for instance, "the national" is usually conceptualized as being above "the urban" but below "the global." Moreover, this verticality is often equated with a sense of significant power, such that those who are viewed as being higher up the scalar hierarchy (such as those considered "national" or "global") are seen to enjoy a position of power over those below them (those considered "local" or "regional"), a position which allows them to "loo[k] down from above" (Lefebvre 1973/1976: 88) as if surveiling the Earth in some God's-eye panorama.

However, in the case of the concentric circles metaphor, scale's dimensionality is viewed quite differently, as a horizontal aspect of social or natural systems. In this metaphor, each scale is seen as part of a series of concentric circles stretching out from a central circle. Typically, the outer circle is conceptualized as "the global" scale, whereas the innermost circle is seen as "the body," "the local," or "the urban," depending upon the specific smallest scale considered in any particular set of scales. Other scales, such as "the regional" and "the national" lie between these two scalar extremes. Certainly, this metaphor shares some important similarities with that of the ladder. Thus, each circle is seen as a separate scale in the same way as is each rung of the ladder. Moreover, there is also typically a sense of greater or lesser power associated with specific scales – the outermost scale, for instance, being the biggest, is generally considered the most powerful. However, there are also some significant differences between the two metaphors. In particular, whereas in the ladder metaphor "the global" is habitually considered to be the "highest scale," one that sits above others, in the concentric circles metaphor it is customarily regarded as the "largest scale" or the "outermost scale." Rather than a language of above/below, the concentric circles metaphor draws upon a language of enclosure and/or encompassing, with scales enclosing/encompassing, or being enclosed/encompassed by, others – "the global" encompasses all other scales, whilst "the national" scale may both encompass "smaller" scales such as "the urban" but itself be encompassed by "the global." There are, then, in these two metaphors two understandings of the relationship between various scales which have some overlap but which are also quite different in some fundamental ways.

These two metaphors, however, are not the only ones that have been used to present the world as scaled. In particular, the image of a tightly nested hierarchy using the metaphor of Russian Matryoshka ("nesting") dolls has also been quite commonly utilized (see Figure 1.3). This representation shares some commonalities with both the ladder and the concentric circles metaphor – each doll/scale is clearly identifiable as a separate and discrete entity. Equally, the outside doll is the "largest" scale which contains all others (as with the concentric circles metaphor), though it is not really "above" other scales in the same manner as in the ladder metaphor. However, probably what most distinguishes this metaphor is that each individual scale fits together as part of a whole in a much more rigidly conceived progression than in either the concentric circle or ladder metaphor. Indeed, the scalar gestalt is only really complete if and when each doll/scale sits inside the one that is immediately bigger than itself. This has three consequences.

First, all of the dolls/scales fit together in one and only one way, for a larger doll/scale simply will not fit inside a smaller one. There is no possibility of representing, say, "the local" as containing "the global" in the manner in which some theorists have suggested might occur à la glocalization. Second, whereas with the ladder or circles metaphors it is possible to conceive of certain scales being skipped as social actors move from, say, "the regional" to "the global" scale, this is less the case with the Matryoshka metaphor. Hence, although in the ladder metaphor we could, perhaps, imagine a particularly agile company jumping from a

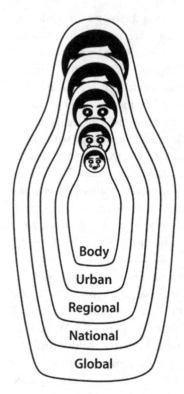

Figure 1.3 Scale as nesting dolls

"regional" geography of organization to a "global" one by leaping over "the national" rung/scale, it is much more difficult to do so when conceiving of scales as highly integrated Matryoshka dolls. Finally, the Matryoshka metaphor leads, perhaps, to yet a third set of discourses. Consequently, if the ladder metaphor generates a language of "above/ below," and the concentric circles metaphor a language of "larger/smaller" and/or "encircling/encircled," we might conceive of how the Matryoshka metaphor may promulgate a language of "containing/contained" (given that circles are two-dimensional objects but Matryoshka dolls are three-dimensional).

Whereas the ladder, concentric circles, and Matryoshka dolls metaphors all view scales in areal terms, as we have seen, the image of the network has also become very popular in the past decade or so, the result, perhaps, of the growth of computer technology. Hence, Latour

(1996: 370) has argued that the world's complexity cannot be captured by "notions of levels, layers, territories, [and] spheres," and so should not be thought of as being made up of discrete areas of bounded space which fit together neatly. Rather than seeing the world as constituted by hierarchically ordered "space envelopes," Latour maintained that we must think of it as "fibrous, thread-like, wiry, stringy, ropy, [and] capillary." Such suggestions that scales are more appropriately thought of in rhizomic than in areal terms present opportunities for quite different sets of metaphors. For instance, one metaphor to which such a view lends itself is that of scales as a series of tree roots (see Figure 1.4). In this metaphor, scales are not viewed as separate spatial arenas but as

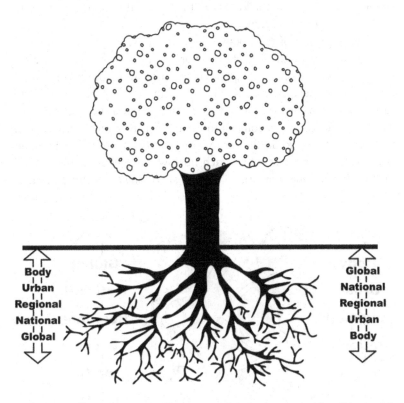

Figure 1.4 Scale as tree roots. The tree trunk can represent either "the body" (or perhaps "the local"), the place where more widespread global processes surface/come together in a single point, or "the global," the point at which myriad deeper local, regional, and national scales come together

locations along various parts of networks, with scales such as "the global" and "the body" or "the local" presented not as opposite ends of some scalar spectrum but as a terminology for distinguishing shorter and less-connected networks from longer and more-connected ones. What is particularly interesting about such a metaphor is that the scalar hierarchy which is frequently seen to place "the global" at one end of a spectrum against some Other ("the local," "the body," "the urban") at its opposite end can be understood to run in two directions. Thus, on the one hand, the point where the tree trunk emerges from the ground can readily be thought of as representing "the global" scale, with its roots reaching deep into the soil of other scales. However, this same point could just as easily be seen as the scale of "the local" or "the body," the starting point from which all other scales originate.

If the metaphor of tree roots represents one vision of scale drawing upon a network motif, that of earthworm burrows represents a second (see Figure 1.5). Such a metaphor shares many similarities with that of the tree root metaphor. Thus, much like the tree's roots extend into various layers of the soil, thereby expressing the idea that social and/or natural processes or phenomena may run deeper or more shallowly in various situations, so do the earthworms' burrows. Likewise, the point at the Earth's surface at which each burrow emerges from the ground can be taken to represent either "the global" or one of its Others. Hence,

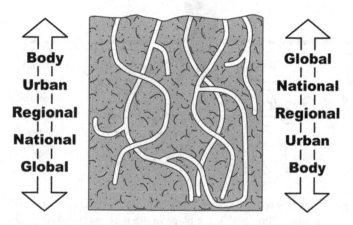

Figure 1.5 Scale as earthworm burrows. As with the "scale as tree roots" metaphor, the depth of the burrows in the soil or their openness at the Earth's surface can be read in different scalar ways

if "the global" is represented as being closest to the Earth's surface, its Others are seen to run deeper into the soil/society/Nature – they are quite literally more grounded. Contrarily, this metaphor could just as easily represent "the global" scale as stretching deepest into the soil, with its Other being closest to the Earth's surface – the surface is, perhaps, the point at which deep global processes and phenomena become visible to the eye. Moreover, for both the tree root and the earthworm burrow metaphors, scales are portrayed not as separate from one another but as linked together in a single, interconnected whole, with the result that whilst different scales may be recognized (much as one can think of tree roots or earthworm burrows breaking through different soil strata, with some going deeper and some being more shallow), it is tricky to establish precisely where one ends and another commences. Additionally, whereas the ladder and circle metaphors represent two-dimensional understandings of the world, the tree root and earthwork burrow metaphors both represent three-dimensional ones (even if they are presented graphically in two dimensions on paper). However, there is also at least one fundamental difference between them: whereas the tree root metaphor generally privileges a single point of entry into the scaled world (the point at which the tree's trunk penetrates the ground), the earthworm burrow metaphor can suggest multiple points of entry into this world, given that such a burrow system may have several tunnel openings.

A third metaphor that draws upon the idea of networks is that of the spider's web, which is often taken to denote non-linearity, non-hierarchization, decentralization, and the criss-crossing of linkages between various parts of the Earth's surface. In such a metaphor, a scale such as "the local" may be seen to sit at the center of the web, connected to a set of supra-local points across the network, or, alternatively, "the global" may be seen to sit at the center, linked to numerous local places by the web's threads but serving as its focalpoint. As with the previous two network-inspired metaphors, the spider's web sees the world as highly interconnected and, recalling Latour, as fibrous, thread-like, wiry, stringy, ropy, and capillary. At the same time, though, the web metaphor is, in important ways, quite dissimilar from these other two. Thus, whilst neither the tree root nor the earthworm burrow metaphors suggest necessarily that spaces are enclosed, in the case of the spider's web spaces within it are, in fact, bordered in much the same way as they are in areal views of scales (such as the concentric circles metaphor), though with one crucial difference: the spaces so bordered are, paradoxically, a sort

Figure 1.6 Scale as a spider's web, with trapezoids of empty space enclosed by the web's threads. In such a web, scales are not represented by areal hierarchies but by connections between different nodes/points in the web, such that shorter connections might be considered to represent, say, "regional" links whereas longer ones might be considered to represent "global" links

of non-space – they are spaces of absence rather than of presence. In approaches that view scales in areal terms, the spatial boundaries which delineate and enclose various absolute spaces are integrally connected to them and those spaces are viewed as having substance (they are, after all, seen as parts of the landscape). In the case of the spider's web, on the other hand, the spaces between the threads are not actually part of it. (One might think of the holes in a lump of Swiss cheese in similar fashion – they are both part of the cheese but also not part of it.) Rather, the mass of the web is made up of only the silken threads which provide its form, whilst none of the trapezoids outlined in each segment of the web – trapezoids that are fundamental to giving the web form – themselves have mass (see Figure 1.6).

Taking this metaphor's exploration a little further, whilst in portrayals of networks as webs it is often suggested that "[n]o central node sits in the middle of the ... web, controlling and monitoring every link and node [, and t]here is no single node whose removal could break the web" (Barabási 2002: 221), a fact which reinforces the web's decentralized image, viewed from a different perspective the decentralization and non-linearity proclaimed by the metaphor disappear and are replaced by centrality and linearity. Hence, in the material – as opposed to the metaphorical – world, the spider, the web's solitary producer, typically sits at the web's center surveiling its farthest reaches awaiting an unfortunate insect to fly into any part of its domain, whilst the web itself is spatially defined by the linear and quite geometric threads which constitute it. Moreover, although the spider's web is often presented as an image that epitomizes an entity with dimensions but not with scalar hierarchy, networks themselves do not necessarily lack hierarchy. Thus, although some natural and social scientists have adopted Barabási's (p. 221) suggestion that it is possible to have a "scale-free network," with the latter being seen as "a web without a spider" (i.e., a web without a central controlling entity), as Barabási himself has pointed out, even though what is arguably one of the most decentralized networked organizations in the contemporary world (al Qaeda) may well have evolved over several years "into a self-organized spiderless web," it nevertheless contains "a hierarchy of hubs [which have] kept the organization together" (p. 222). Furthermore, although Barabási offers that such a decentralized, spiderless network such as al Qaeda may have developed in an organic manner and that, in "the absence of a spider, there is no meticulous design" (p. 222), in the material world a web cannot come into existence without the spider's existence – the spider's web metaphor, then, only goes so far.

There are other similarities and differences between the web metaphor and some of the others that have been used to describe networks' structure. Hence, on a fully formed spider's web it is possible to get from any one point on the web to any other point by simply following various threads as they intersect each other – no node, in other words, serves as the central point through which one must pass to traverse the network/web. Although this shares some similarities with the earthworm burrow metaphor, where one might be able to get to one point of the burrow complex from several other points without having to go through a central node, it is quite different from that of the tree roots, where crossing the network will generally require passing through a central point (depicted as the point where the tree trunk leaves the ground).

Equally, although the spider's web is a powerful metaphor, it is important to recognize that the generally symmetrical image of a wheel-shaped
web presented in such a metaphor to stand in for all spiders' webs is, in
fact, quite particular to the orb-weaver (*Araneidae*) spiders, and that other
types of spiders make webs with quite different structures, as with the
funnel-weaver (*Agelenidae*) spiders or the neotropical spiders who
communally – rather than individually – make webs which may stretch
several miles. Moreover, although the *Araneidae* spiders' webs are typically
graphically presented as symmetrical, in actuality they generally exhibit
a degree of north/south asymmetry, the result of gravity (Witt *et al.*
1977). Finally, whereas earthworm burrows and tree roots are actually
three-dimensional in reality (even if portrayed two-dimensionally on
paper), webs produced by orb-weavers are generally two-dimensional
in form, whilst those of other spiders may be three-dimensional.

If the ladder, concentric circles, and Matryoshka dolls metaphors all
represent different ways of portraying scales conceived in areal terms,
and the tree root, earthworm burrows, and spider's web metaphors
all represent different ways of portraying scales conceived in network
terms, then another, radically different, metaphor is that of musical
scales (Howitt 1998). Outlining how different musical scales incorporate
different notes in different ways, even as those notes themselves remain
constant, Howitt highlighted how this metaphor facilitates a more
relational view of geographical phenomena. Thus, he indicated that the
musical metaphor would allow one to "see that in a geographical totality,
many elements will remain consistent in a geographical analysis that
spans across different geographical scales" and that what changes in
such analysis "is not the elements themselves (the features on a landscape,
the sites involved in a production process, the ecological processes
affecting a social formation, the cultural practices performed by people),
but the relationships that we perceive between them and the ways in
which we might emphasize specific elements for analytical attention"
(p. 55). Using the example of a bauxite mine at Weipa, Queensland
(Australia), Howitt proposed that its relationship to particular scales ("the
regional," "the national," "the global") was quite different and changed,
depending upon the specific scale about which one is talking, "in a way
that is similar to the way in which we might find a C note in several
scales playing quite different roles in the musical totality, even though
neither the C note nor the Weipa mine changes as a material phenomenon
in these different scale contexts." For Howitt (p. 56), adopting a musical

metaphor facilitates "a shift in understanding of scale from an (over)-emphasis on scale as size and/or . . . level, to include aspects of scale as relation." Such a shift, he averred, allows a consideration of "not just the sorts of connections (relations) that help to constitute particular geographical scales, but also to begin to see geographical scale as . . . a factor in itself, a structure, system or unit that can be abstracted from geographical totalities as having some relatively autonomous (though never independent) causal efficacy." The result, he maintained, was to provide an approach that "may provide a better way for us to talk about why scale is a co-equal concept with more theorized notions such as environment, place and space."

Certainly, Howitt's approach provides a quite different metaphor for understanding spatial scales than do the others addressed above. It also shares similarities with some recent scale theorizing (the idea that "scale is a co-equal concept," rather than the dominant one, mirrors claims by Leitner *et al.* and Jessop *et al.*, amongst others) but also some differences (the idea that scale might be considered "a factor in itself" contradicts arguments that scale should be viewed dialectically). At the same time, as with the other metaphors explored above, it is important to recognize the limitations with this one. Hence, users of this metaphor must remember that the musical scales which are popular in Western musical traditions are not necessarily those of other cultures. For instance, whereas in traditional Western music scale notes are frequently separated by semitones (a half-step between notes) to create twelve pitches per octave, musical traditions from other parts of the globe utilize scales that have other intervals and/or a different number of pitches. Equally, in some other musical traditions (as in that of India) some notes are produced that cannot be reproduced on the standard European piano. This, however, is not to say that the musical metaphor is not useful for understanding scale. Rather, it is to point out that, as with taking the orb-weaver's web to be representative of all webs when, in fact, using a different type of spider's web as an exemplar may change that particular metaphor's meaning, it is important to recognize that drawing upon diverse musical traditions from different parts of the world may change how we understand scale through this particular metaphor.

In surveying some of the metaphors used to conceptualize geographical scale, my goal has not been to provide an exhaustive accounting of all of the metaphors that have been – or could be – used when representing how the world may or may not be scaled. Instead, what I have tried to do is to illustrate how scales and their interrelations have been represented figuratively. In so doing, it is important, again, to stress that using different metaphors to talk about scaled relationships between places is not to suggest that these various metaphors inevitably represent different states of affairs in the material world, nor that one is necessarily a better representation of the world than is another. Rather, metaphors are of great consequence because they shape how we conceptualize and interact with the world and, thus, how we make geographical landscapes.

CONCLUDING REMARKS

In this opening chapter, I have sought to address two issues concerning scale. First, I have tried to show how different analytical traditions have conceived of scale and the scaled (or not!) nature of the world. Second, I have attempted to show that different scalar metaphors bring with them different consequences for how we conceive of, and so engage with, the world. Three interconnected issues, however, emerge out of this narrative.

1)   In considering how the debate over scale has progressed, the recent turn toward semiotics and post-structuralism appears to have heralded the emergence of a neo-Kantianism when conceptualizing scale. Thus, much discussion has focused upon epistemological rather than onto-logical questions – it has focused upon how scales are thought of and how this shapes consciousness rather than on how scales as material entities structure, and are structured by, economic, political, and environ-mental processes. Certainly, this is not to say that the world of ideas is unimportant. Clearly, how we conceive of scales does shape in important ways how we behave in the world. But it is to say that the materiality of scale should not be forgotten.

2)   One of the key issues concerning any analysis of scale is what Smith (1987) called a "Gestalt of Scale" – that is to say, the way in which different scales fit together to form an overall pattern and how looking at them from different perspectives can result in very different under-standings of material reality. Thus, using the example of suburbanization, Smith contended that where one stands determines what one sees with regard to processes of rescaling. Hence, from one perspective, the "growth of the suburbs is generally treated as a spatial decentralization

of the city, and indeed from the office towers or from city hall located at the centre, suburban growth is a decentralization outward." However, from a slightly different perspective – say, observing the city from a hot-air balloon floating over it, a position from where one can see across the broader urban region – "the rapid growth of the suburbs now seems like an impressive *centralization* of the region's economic activity." Whether one understands suburbanization to be a process of decentralization or centralization, both or neither depends upon where one is standing at any particular time, for although "the empirical pieces are the same . . . with a slight shift of perspective they make a very different pattern . . . If you view the pieces from one scale you see one pattern (or lack of pattern), and if you view it from another scale you see a different one." The danger in all of this for Smith (and other materialists) is that scale may be presented as little more than an optical illusion, a kaleido-scopic effect. However, recognizing that how one sees scale may change depending upon one's geographical position is a different matter from saying that changing one's geographical position changes how scales are constituted materially and how they function.

3)   The relationship between the ideational and the material raises key questions concerning how narratives about scale function, for the ways in which narratives are structured can lead us to understand material practices – such as scale production – in particular ways. Thus, as Sayer (1989: 263) argued, "the depiction of events chronologically, in a story, gives the appearance of a causal chain or logic and the sense of movement towards a conclusion" where, in fact, there might be none. Likewise, "not only writing but research itself is influenced by its location within ongoing debates and against rival academic and political groupings," such that understandings of material processes and discursive representa-tions will be shaped by the understandings that have come before them. The result is that there can be a certain "autonomy" to narratives, such that the discursive structures shaping them may or may not mirror those which shape the material processes they are supposed to be signifying. Hence, whereas the emergence of a particular region – Sayer gives the example of Silicon Valley – is often presented in quasi-teleological fashion as if it were an inevitability once a few initial catalysts were put in place, in fact myriad events – some causally connected, others purely happenstance – may have been required. Consequently, when describing how regions have come into being over historical time, it is important

to be able to distinguish whether causal relations have been at work or whether the region is simply seen to have emerged historically because of the way in which the narrative has been laid out. The key is to comprehend how the structure of a textual narrative relates to actual historical and geographical processes that may be involved in creating regions. This means that it is important to excavate the connections between representations of scales and how scales function in the material world, tricky as that may be, so that changes in the former are not taken to indicate changes in the latter whilst changes in the latter may be seen to be reflected in the former (Smith and Katz 1993). The narrative form also means that social products which are inextricably connected materially may be represented as separate entities because of the necessity of describing them serially. Frankly, this is an issue in what follows in this book – although the various scales which I will address in the following pages are necessarily connected (for instance, the forces of global capitalism shape the human body, whilst myriad human bodies are brought together to drive global processes), what Sayer (p. 270) calls the "tyranny" of the sequential text means that they are addressed separately. It is important to recognize, however, that representation is not necessarily reality and that their depiction as separate entities does not mean that they are in fact so.

2

THE BODY

Some scientists find it helpful to think of the human body as having a hypothetical outer boundary that separates the inside of the body from the outside. The outer boundary of the body consists of the skin and openings into the body, such as the mouth, nostrils, or punctures and lesions in the skin.

Hayes (2001: 390)

The body is man's first and most natural instrument.

Mauss (1936/1979)

In the electric age . . . our central nervous system is technologically extended to involve in the whole of mankind and to incorporate the whole of mankind in us.

McLuhan (1964/2001: 4–5)

The body has frequently been seen within the humanities and social science literatures as the scale that, in some ways, serves as the foundation upon which all other scales are based. We can, perhaps, think of several reasons for this: bodies are often thought to be easily identifiable as individual entities and can thus be seen to serve either as discrete social building blocks of larger units (families, communities, nations) or as nodes within broader networks that link places together; bodies possess particular boundaries (i.e., the skin) which enclose them and separate their interior spaces from the broader world beyond; and bodies have been a particular focus of much critical theorizing. Significantly, the body as a scale is also very different from other scales of social organization such as the global or the national, for whereas these other scales are socially produced, bodies are both biological entities and social ones, with one's biological shape and size being very much shaped by social conditions – to take an obvious example, not having sufficient money to buy food (a social condition) can have severe detrimental effects on one's metabolism, either in the short or the long term (a biological condition). This fact has bearing upon considering "the body" as a scale.

In this chapter, I will explore recent debates about the body as they relate to issues of geographical scale, for how we think about "the body" as a body shapes how we think about it as a spatial scale. First, I will explore some of the ways in which the body itself has been theorized as an entity. Second, I will examine several issues to consider when talking of the body as a scale, for despite the significant theorizing of bodies qua bodies, there has been little critical consideration of how such theorizing might also be applied to thinking about the body qua scale – for instance, within the scale literature the body has often been seen to be a rather unproblematic, self-evident thing. Finally, I consider what are some of the consequences for thinking about different understandings of scale – scale as areal versus as networked, for instance – which flow from how the body qua body has been theorized.

THEORIZING THE BODY

Within the social sciences in the past two decades or so, it is probably fair to say that feminist theorists have been the leading voices contemplating the body's status and position within society, if from several different perspectives. Some strands of theorizing have viewed the body from a more biological point of view, essentializing it and seeing differences in women's and men's social positions as resulting from their diverse biologies. Hence, whereas many non-feminist writers had historically argued that women were socially and physically inferior to men because of their biology, some feminists adopted the position that biology was indeed a determiner of social position but simply sought to turn things on their head – rather than men being considered superior, they considered women to be so because of, for instance, their ability to give birth and consequently their allegedly "naturally more nurturing" demeanor. Others, whilst not going so far as to argue for women's superiority, averred that women were naturally better able to understand particular aspects of society because of their social position within it – a stance often referred to as "standpoint feminism" (Hartsock 1983). Griffin (1978), for instance, suggested that women's biology gave them certain insights – especially regarding Nature – that men could not have, whilst Firestone (1970: 8–9) developed an analysis in which she argued that "biology itself – procreation – is at the origin of the dualism" between men's and women's social roles in society, suggesting that "throughout history . . . [women] were at the continual mercy of their biology . . . which made them dependent on males . . . for physical survival." However, she argued, the development of modern technologies allowing reproductive choice now allows women to escape their biology – biological determinism, in other words, has historical limits. Taking a slightly different tack, still other feminist writers (e.g., Foord and Gregson 1986) have maintained that the facts of biological reproduction place men and women in different sets of relationships within capitalism (and other social systems), such that patriarchy cannot be avoided – even if its forms take on different characteristics in different times and places – and that, consequently, there seems to be something of a universal female experience, one based ultimately on biology.

Whereas one strand of theorizing seemed to essentialize the body as a natural base upon which social relations were developed, other feminists have sought to get beyond such a biological basis for understanding women's oppression. Hence, de Beauvoir (1949/1953: 267) famously declared that "[o]ne is not born, but rather becomes, a woman. No biological, psychological, or economic fate determines the figure that the human female presents in society; it is civilization as a whole that produces this creature, intermediate between male and eunuch, which is described as feminine." With this turn of phrase, de Beauvoir sought to indicate that there was an important distinction between sex (determined by biology) and gender (determined by society). Specifically, her purpose was to explore the gender binary which has generally seen women as Other to men, who are perceived as the standard for humanity against which women should be measured. In her work, de Beauvoir challenged the idea that women are abnormal and inferior to men. However, in trying to do so she has often been accused of simply reproducing the masculine/feminine binary in a different form by accepting that masculinity is the norm to which femininity should aspire and by not considering much the possibility that genders other than masculine and feminine might exist – she views lesbians, for instance, largely in terms of what she considers heterosexual male and female behavioral traits. Moreover, in distinguishing between sex and gender, de Beauvoir paradoxically appeared merely to reinforce another binary (sex/gender), one in which biology still served as a kind of "coatrack" for gender – in such a dualism the body is largely "viewed as a type of rack upon which differing cultural artifacts, specifically those of personality and behavior, are thrown or superimposed" (Nicholson 1994: 81).

If some feminists have explored how social relations might evolve out of biological attributes, other scholars, rather than viewing the human body as an entity shaped in a unidirectional manner by biological evolution, such that we are solely prisoners of our biology, have maintained that humans have, in fact, played an active role in shaping our own biological evolution over time. Hence, Tanner (1981) has argued that sex selection of mates by female primates historically shaped which genes were passed on to the next generation, whilst modern primatology shows that female chimpanzees tend to use tools more frequently than do males for food gathering. Assuming that human ancestors did likewise, Tanner suggests that tool use had an impact upon the amount of food available to early hominids which, in turn, allowed a longer period of dependency of offspring upon parents and, thus, shaped our evolutionary path. Moreover, it may even have played a role in the development of human thinking and communication skills – as our ancestors' food sources changed as a result of tool use, the large molars typical of other primates became less advantageous, which left more cranial capacity for the brain and voice box (Eisler 2000). Others have similarly suggested that early humans' use of tools shaped their subsequent evolution. Haraway (1989: 208), for instance, has argued that "Man is his own product," a fact which has important consequences for the social production of the body as a scale.

Whereas one strand of feminism has seen the body largely as a natural entity out of whose biology gender roles emerge, and another has seen the body's biology as at least partially socially produced through humans' own actions, a third strand – one generally associated with postmodern and post-structuralist thinking – has sought to move beyond the realm of the biological to that of the categorical, that is to say the realm of how bodies are constructed as social objects. For example, Bordo (1993) suggested that the body should not be thought of so much as a biological entity but as a text upon which cultural meanings are projected. In this regard, the body is "an effect not of genetics but of relations of power" (Davis 2007: 127). For her part, Butler (1990) has challenged the biology/social dualism through her ideas concerning performativity, namely that gender is "simply" performed through ritualized actions of various subjects. Rejecting both the biological essentialism of some and the "body as text" of others, Butler (p. 136) contended that "the gendered body . . . has no ontological status apart from the various acts which constitute its reality" and that, rather than there being an inner biological core to bodies upon whose exterior surfaces cultural meanings are projected, "acts and gestures, articulated and enacted desires create the illusion of an interior and organizing gender core, an illusion discursively maintained for the purposes of the regulation of sexuality within the obligatory frame of reproductive heterosexuality." Finally, in her analysis of workers in Mexican *maquila* plants, Wright (2001: 356) explored the idea of the body as hermaphroditic, not in a biological sense but in a social sense. Contending that within these factories female assembly workers are trained to internalize their male supervisors' thoughts, she projected a vision in which such supervisors "are constituted as flexibly trained cerebrums" and are paired with the "*untrainable* digits, eyes, and limbs of female operators," such that out of "the coupling of these opposed and partial figures . . . a third hermaphroditic body emerges," one whose head is male but whose hands and arms are female. This hermaphroditic body is not simply the result of melding the female into the male but, instead, "forces them to work together in the forming of a supervisory subject, [a subject] which is greater than the sum of its parts."

As Davis (2007) has detailed, Haraway (1991) took a slightly different approach to overcoming the biology/culture binary through developing the notion of the cyborg, wherein rigid boundaries, such as those separating humans from animals and humans from machines, are challenged (I shall return to the issue of the cyborg below). However, in response to the belief that Haraway's approach presented "a decidedly ethereal body" (Davis 2007: 128), one far removed from the material body, writers such as Grosz (1994) developed the concept of the "volatile body," which is capable of incorporating external objects into its own internal spaces in a continuous process of becoming. The result was that in "stressing the fluidity of bodily boundaries and the dynamic interface between the internal and the external, between the body and its surroundings, [Grosz] attempted to salvage the flesh and blood body while countering any suggestion that biology was destiny" (Davis 2007: 128). Nevertheless, for some such an approach has resulted in – paradoxically – a disembodied body, in which its biological/material aspects have disappeared and the body has been left as simply a metaphorical construction, one which focuses upon the body's surface rather than its sensual physicality. Such feminist body theory, according to Davis (2007: 129), "implies a transformability that belies the bodily constraints with which most women must live at different periods in their lives . . . [and which have] little to offer when it comes to understanding the vulnerabilities and limitations of the body that accompan[y] illness, disease, disability, or the vicissitudes of growing older." In a slightly different vein, writers such as Bordo (1987) have outlined how men are often conceptualized as capable of escaping the confines of their bodies and viewing the world from a universal or God's-eye position whereas women are seen as prisoners of their biology – the male body, in other words, is transcendent whereas the female is fixed.

Although they do not necessarily draw upon feminist theory to do so, following from the above it is important to recognize that a number of geographers (and others) have explored the body in terms of people's differential abilities to cross space and/or to challenge spatial barriers. Whereas Rowles (1978) explored how the aged body is spatialized, and Katz (2004) how the juvenile is so, yet other writers have examined the disabled body in space. Whilst some of these have investigated how disabled people have engaged with the state, creating new scales of organization in the process (see Kitchin and Wilton 2003), others, for instance Davis (1995: 23), have considered more deeply the disabled body as a social construction to survey how "disabled" bodies are disciplined through efforts to portray them as abnormal, suggesting that "[t]o understand the disabled body, one must return to the concept of the norm, the normal body." Thus, drawing a parallel with studies of race, Davis (1995: 23–24) has argued that much writing about disability "has focused on the disabled person as the object of study, just as the study of race has focused on the person of color," and has suggested that "the 'problem' is not the person with the disabilities; the problem is the way that normalcy is constructed to create the 'problem' of the disabled person." Others, meanwhile, have sought to draw distinctions between "disability" and "impairment," seeing disability as a collective issue and impairment a personal one, although this approach reproduces modernist dualisms (disability/collectivity v. impairment/individuality) which postmodernism has been keen to overcome. Thus, as Hughes (2004: 66) has maintained, whereas writers steeped in the traditions of modernism have tended to sustain a worldview in which they see real biological differences between disabled and non-disabled bodies, postmodernists have generally seen the differences between abled and disabled as culturally – rather than biologically – based. Hence, he avers, whereas the non-disabled body is usually described as "normal," such a "normal" body is itself not real in any ontological sense but is, rather, merely a statistical average of billions of human beings' corporeal morphologies. Consequently, if the "normal" body is really a socially constructed object, how can the "non-normal" body be anything but likewise?

Such ideas of how a bodily norm is established have also been probed by critical race and queer theorists. For their part, critical race theorists have challenged theories which see race in biological terms and racial categories as stark and fixed things (see Brown 1998). Omi and Winant (1994: 55), for instance, have shown how "race" is neither an essence with regard to the human body nor an illusion but, rather, an "unstable and 'decentered' complex of social meanings constantly being transformed by political struggle," struggle which allowed, for instance, the Irish, Italians, and others in the United States to "become white" (Ignatiev 1995) in the decades before the First World War. Moreover, such theorists have maintained that within Western culture – and perhaps more broadly – "whiteness" has generally been taken to be the norm within a binary against which other things are measured (as in the oft-used descriptions of people as either "white" or "non-white"). Hence, critical race theorists have suggested that even though people from different parts of the world obviously look different, the racialized body is not so much born as it is created – "races" simply do not make much sense as biological entities (Cooper 1984; Brown 1998). Furthermore, within the discourses of the racialized body as they have been expressed in Western societies, it has generally been the case that certain racialized bodies have been seen to enjoy privileged positions in different places. Hence, generally the white body is seen as dominant but the non-white one may be seen to offer a unique voice because of its positionality within a white-dominated society – a view emblematic of the so-called "voice-of-color thesis" (Delgado and Stefancic 2001: 9).

Queer theorists have likewise questioned the body's status and how it fits into particular dualisms, in this case its presumed heterosexuality and beliefs that all bodies are imbricated within an asymmetrical male/female binary. For instance, whilst early feminist work in the social sciences questioned the role typically ascribed to women, it tended to do so within the context of an assumption of heterosexuality as normal. However, beginning with the work of writers such as Rich (1980: 632), who argued that in such feminist works the "lesbian experience is perceived on a scale ranging from deviant to abhorrent, or simply rendered invisible," such heteronormativity (the assumption that heterosexuality is normal and, therefore, that homosexuality is abnormal) came to be questioned. In particular, within the feminist literature there was a clear divide between those who saw all women as having a degree of essentially similar experiences with regard to their relationship to men and those who saw the experiences of heterosexual and homosexual women (and men) as quite different. Complicating the picture further, there arose debates about the privileges that gay men enjoyed relative to lesbian and/or straight women because they were men, even within a society that considered homosexuality deviant.

Such deliberations produced several acknowledgments, in particular that the sexualized binaries – male/female, straight/gay – that had frequently been used to conceptualize the body were insufficiently nuanced. Hence, how might transgendered bodies be understood to transcend such binaries, and how might such socially defined binaries themselves serve as the basis for biologically engineering individuals' bodies, as when surgeons have "corrected" children's ambiguously non-binary sexual organs (as has been done with children born as hermaphrodites)? Equally, how might individuals' sexuality be performed in different ways in different places? Thus, as Brown (2000) has suggested, homosexual men and women who are still "in the closet" – itself a spatial metaphor – may act in different ways both from those who are "out" and depending upon their geographical location (in countries where they may be put to death for engaging in homosexual acts, "being out" carries considerably greater dangers than in places such as San Francisco). Likewise, homosexual – or at least non-heterosexual – bodies face certain challenges in existing in, and crossing through, "heterosexual spaces" (Bell and Valentine 1995; Valentine 1996). In sum, queer theorists destabilized the binaries that had for so long undergirded much writing about the body by suggesting that perhaps a continuum of sexualities might be more appropriately considered. Many geographers working in this area have detailed how sexualities are lived and performed differently in different social and spatial contexts.

Whilst feminism, critical race theory, and queer theory have all played important roles in theorizing the body within Geography and other social sciences, another important strand of theorizing has been that which has emerged out of the psychoanalytical literature (Pile 1996; Kingsbury 2004). Importantly for our consideration of the body as a scale, much of this writing has both employed a binary of inside/outside and has its roots in biological explanations of human behavior. For instance, Sigmund Freud famously declared that "anatomy is destiny," suggesting that morphological differences between the sexes expressed themselves in differences of the mind and that there is thus a biology/culture binary at play in the formation of human sexual consciousness. Furthermore, within Geography – particularly the Behavioral Geography of the 1970s – much of the work influenced by psychoanalytical approaches has viewed the body as a container for the mind and focused upon how the mind interacts with the environment in which the body containing it lives. Such a body–mind dualism was explored by Descartes in his 1641 *Meditations on First Philosophy*, wherein he maintained that the material body and the immaterial mind were ontologically separate things and that the body was the mind's container, although the two nevertheless causally interacted. Others have also used binaries to explore either the mind or the mind/body relationship. Hence, Freud sought to discover the link between the conscious mind and the unconscious one (he called (1927: 9) the division between the two "the fundamental premise on which psycho-analysis is based"), whereas Foucault, particularly in his analysis of the history of crime and the birth of the modern prison system (Foucault 1975/1977), suggested that in the late eighteenth century the focus of punishment in the West began shifting from the corporeal body (which had often been disciplined up to that point by, say, being beaten with whips or branded with irons) to the mind/soul (used here in a non-religious sense) wherein punishment was designed to deprive a body of various liberties and to provide it time in prison to consider the error of its ways. Whereas Plato had argued that the body is a prison for the soul, Foucault suggested instead that the body is a prisoner of the soul. For Foucault, therefore, the body is the focal point of social regulatory technologies, the entity upon which power and knowledge are brought to bear for purposes of producing a disciplined (i.e., socially controlled) citizen. Hence, for him the body is an object to be controlled publicly by external forces – those of the prison or the asylum, for instance – whilst the soul is a private entity, judged to be either normal or abnormal and, based upon this, to be in need of reform or not. Moreover, Foucault developed the idea of the body as a "site," in which its "normalization" depended upon its spatial situation – as he (1975/1977: 143) put it, "[e]ach individual has his own place; and each place its individual," a position that somewhat mirrors Lefebvre's (1974/1991: 170) declaration that "each living body is space and has its space; it produces itself in space and it also produces that space."

Significantly, in much of this psychoanalytical writing there is a binary language of inside/outside, of bodily surfaces and bodily depth. Hence, Freud (1927: 31) argued that "the ego . . . is not merely a surface entity, but . . . is itself the projection of a surface." Likewise, poet and literary critic Susan Stewart (1984: 104) conceptualized the body in quite geographical terms as simultaneously "contained and container," suggesting that such a conceptualization means that "our attention is continually focused upon the boundaries or limits of the body; known from an exterior, the limits of the body as object; known from an interior, the limits of its physical extension into space." Equally, for Lacan (1977) the body's surface serves as a divider between its innerworld and outerworld, such that apertures which transgress it (the mouth, ears, anus, etc.) take on significant cultural import, as do the materials which cross through such apertures (feces, urine, sweat, food and drink, and blood, both menstrual and non-menstrual). Consequently, as Falk (1994: 3) suggests, the body is often seen as something which the outside world disciplines but also as providing the point from which an individual defines "his or her (bodily) boundaries and the relationship to the 'not-me'" – that is to say, the body's physical limits are seen to define the boundary between the Self and its Other, between the subject and the objects of the outside world. Hence, "[t]he expression 'me and my body' [is] a distinction in which the body is defined as something outside 'my self' – as an out-look and/or an object to be molded, disciplined, cultivated, etc. and as an instrument used in material or symbolic (expressive) 'labour'" (Falk 1994: 2–3). This distinction is culturally defined – as Bloustien (2001: 100) maintains, "in order . . . to conceptualise [the] body . . . in terms of 'inside' and 'outside' . . . [it is necessary to draw] on an historically and culturally specific mental process that defines 'me' in relation to 'not-me'."

In considering such matters, it is important to recognize that poststructuralist thinking has sought to break down such dualisms – for instance, by conceiving of the body as a layered system in which, rather than there being a distinct break between the external cultural/social world and the mind, a break in which subjectivity is viewed "as a psychic rewriting of the physical" (à la Freud and Lacan) or the body is seen "as a surface for social inscription of psychic interiority" (à la Foucault) (Oikkonen 2004), interior and exterior are part of a continuum. Hence, the phenomenologist Merleau-Ponty (1945/2002) challenged the historical split going back at least to Plato between body and soul by advancing a position based on existence itself, suggesting that one does not *have* a body but that, rather, one *is* a body. For her part, Grosz (1994: xii) suggested that a better model for the body than such a binary was that of the Möbius strip (Box 2.1), which "has the advantage of showing the inflection of mind into body and body into mind, the ways in which, through a kind of twisting or inversion, one side becomes the other [and which] provides a way of problematizing and rethinking the relations between the inside and the outside of the subject, its psychical interior and its corporeal exterior, by showing not their fundamental identity or reducibility but the torsion of the one into the other, the passage, vector, or uncontrollable drift of the inside into the outside and the outside into the inside."

Whereas feminists, queer theorists, and postmodernists have frequently focused analytical attention upon the body, Marx and Marxists have often been accused of ignoring the body. This is not entirely true. Indeed, Marx explored the body from four major vantage points. First, he saw the human body as the ultimate source of all transformation of the material world and thus as the original source of wealth. There are two important aspects with regard to his understanding of the body here. To begin with, Marx (1867/1976: 173) established a tension between the human body and Nature, suggesting that "Man" "opposes himself to Nature as one of her own forces, setting in motion arms and legs, head and hands, the natural forces of his body, in order to appropriate Nature's productions in a form adapted to his own wants." In the process of so acting "upon external nature and chang[ing] it . . . he simultaneously changes his own nature." In such a configuration, Marx recognized that humans are both part of Nature and somewhat separate from it and can transform it through work – turning, say, trees into tables and chairs. Perhaps more importantly, however, for Marx the body – specifically, the body of the worker – is the source of all surplus value and, hence, of profit. Thus, although in the production process material objects may be transformed with regard to their utility and form –

Box 2.1 A MÖBIUS STRIP

A Möbius strip is a one-sided, one-edged surface which looks like a twisted, three-dimensional figure of eight. It is named after the German mathematician August Ferdinand Möbius (1790–1868) and is constructed by holding one end of a rectangle fixed, rotating the opposite end through 180 degrees, and joining its end to the first end. There are two types of Möbius strips, the distinction being whether the direction of the half-twist they contain runs clockwise or anticlockwise. Significantly, the Möbius strip has found some commercial uses. For instance, in 1957 the B.F. Goodrich Company patented a conveyor belt based on the strip and claims that because both its sides – rather than just one – are used as a result of the strip's twist, it lasts twice as long as non-Möbius belts (Pickover 2006). The fact that the strip is a single continuous curve with only one side means that if an ant were to crawl along its length, that ant would eventually arrive back at its starting point without ever having crossed an edge – a feature illustrated in a famous drawing by the Dutch artist M.C. Escher (see Figure 2.1). Such properties have led, amongst other things, to at least one (bad!) joke: Q: Why did the chicken cross the Möbius strip? A: To get to the same side.

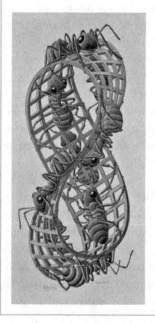

Figure 2.1
M.C. Escher's "Möbius Strip II" © 2009
The M.C. Escher Company-Holland.
All rights reserved. www.mcescher.com

a lump of wood becomes a table – it is only through the application of human labor that an object's value is transformed.

Second, Marx argued that the human body is dramatically impacted by the conditions within which it lives, such that metabolism and biology can be significantly shaped by the social relations of work. Thus, he suggested that as much as the over-zealous pursuit of profit in agriculture might exhaust soil, so might its pursuit more broadly also seize "hold of the other vital force of the nation [labor] at its roots" (pp. 348–49), a fact that could be seen in workers' bodily weakness. Indeed, Marx was quite interested in how the height of soldiers recruited into various European armies had decreased during the century or so since industrial capitalism had emerged, the result, he opined, of poorer nutrition (Box 2.2). Drawing back from the impacts of work and social position on the individual human body, Marx (p. 784) also argued that each Mode of Production (capitalism, feudalism, etc.) had its own law of population dynamics, which affected things such as fertility rates. For Marx, capitalist social relations could have significant biological consequences, even to the point where individual bodies may be exterminated – as in his discussion of the impact of the 1848 potato famine upon Irish agricultural laborers (pp. 861–62).

Third, Marx argued that the body of individual humans as workers had become rendered increasingly invisible as the capitalist system had developed historically. Hence, whereas the handcraft production of goods typical in pre-capitalist societies allowed the mark of individual crafts-persons to be seen in commodities, mechanization increasingly made workers simply adjuncts to the production process and little better than machine-minders – as he (p. 548) put it, "[i]n handicrafts and manu-facture, the worker makes use of a tool; in the factory, the machine makes use of him." Moreover, as long-distance trade routes developed and as the social and spatial organization of production became more complex, consumers' ability to know much about the individual workers who toil to produce the commodities they consume was dramatically reduced, resulting in the subsequent fetishization of the commodity – whilst it is relatively easy to know something about who made particular com-modities, and under what conditions, when they are produced in the same village in which their final consumers live, when they are coming to us from across the ocean this is much less the case.

Finally, Marx used the body as a model for understanding how capitalist economies operate. Thus, he compared the circulation of blood

Box 2.2 THE IMPACT OF SOCIAL RELATIONS ON THE HUMAN BODY

It is often assumed that because the body is a biological entity, it is free from the effects of social relations. This, however, is not the case. For instance, drawing from his observations of workers in Manchester in the 1840s, Friedrich Engels, Karl Marx's compatriot and benefactor, pointed out that the work in which people engaged and the poor diet which they enjoyed (the result of their class position) shaped their bodies' physical development in dramatic ways. Hence, Engels (1844/1993: 252–53) pointed out, minework typically led to:

> the one-sided development of the muscles, so that those especially of the arms, legs, and back, of the shoulders and chest, which are chiefly called into activity in pushing and pulling, attain an uncommonly vigorous development, while all the rest of the body suffers and is atrophied from want of nourishment. More than all else the stature suffers, being stunted and retarded; nearly all miners are short, except those of Leicestershire and Warwickshire, who work under exceptionally favourable conditions. Further, among boys as well as girls, puberty is retarded, among the former often until the eighteenth year . . . This prolongation of the period of childhood is at bottom nothing more than a sign of checked development, which does not fail to bear fruit in later years. Distortions of the legs, knees bent inwards and feet bent outwards, deformities of the spinal column and other malformations, appear the more readily in constitutions thus weakened, in consequence of the almost universally constrained position during work; and they are so frequent that in Yorkshire and Lancashire, as in Northumberland and Durham, the assertion is made by many witnesses, not only by physicians, that a miner may be recognised by his shape among a hundred other persons.

Significantly, Engels's observations are supported by modern statistical techniques. For instance, Komlos (1990: 610) showed that "[c]lass differences in England [were] so great by the end of the eighteenth century that there was a 20-centimeter difference between the height of gentry cadets entering the prestigious Sandhurst Military Academy and the 'Oliver Twists' of the London slums." Similar accounts of how work affects the body biologically can also be found in the contemporary period (Søgaard et al. 2006).

within the human body with the "content-filled circulation of capital" (1858/1973: 517), an approach common amongst other writers of the time – seventeenth-century English political theorist (and physician) John Locke, for instance, had also explicitly equated the circulation of money through the economy with blood's circulation through the body (Finkelstein 2000). At the same time, Marx thought about the human body in rather spectral/supernatural terms. Hence, for him, capitalists were "vampire-like" (1867/1976: 342), economic parasites whose "vampire thirst for the living blood of labour" (p. 367) can never be quenched "while there remains a single muscle, sinew or drop of blood to be exploited" (p. 416) (see Neocleous 2003 for more on Marx's use of allusions to vampirism).

Lastly, one of the most recent writers to address the topic of the body has been Paul Virilio, who has suggested that contemporary technological developments are virtually killing it. One of the things which has most fascinated Virilio is the notion of speed and technology, especially their effect on the body. Thus, he has particularly explored the emergence of what he calls a "dromocracy" – a society of speed – in which the world's wealthy are able to take advantage of the new technologies of high-speed space-crossing (the internet, jet aircraft, high-speed trains) whilst its poor are left behind. Furthermore, as human technology has developed, Virilio (1995a: 100) avers, the human body itself has increasingly been colonized by various technologies (assorted chemicals/drugs, prosthetics, and the accoutrements of modern high-tech life), such that, "[o]ne by one, the perceptive faculties of an individual's body [have been] transferred to machines, or instruments that record images and sound" (Virilio 1993: 4). These technologies, he contends (p. 4), "kill 'present' time by isolating it from its presence *here and now* for the sake of another commutative space that is no longer composed of our 'concrete presence' in the world, but of a 'discrete telepresence' whose enigma remains forever intact." The result is that such technologies are overturning "not only the nature of human environment and its *territorial body*, but also the individual environment and its *animal body*." Consequently, for Virilio (p. 5) the kinds of speed associated with globalization – such as the speed with which commodities, money, information, and people can be moved from one part of the planet to another – are leading to the death of the body as "speed [is] used more and more to *act over distance*, beyond the sphere of influence of the human body and its behavioral biotechnology," and as technology reduces the need for corporeal presence (Smajic 2001).

Clearly, this has implications for the possible erasure of the body as a scale.

BODILY CONSIDERATIONS

Despite the above considerations, within scalar debates bodies qua bodies have generally not been theorized much but have been viewed as fairly self-evidently distinct biological entities, in much the same way that an individual rung at the bottom of a ladder or a circle at the center of a series of concentric circles might be viewed as a discrete and distinct entity. In such a view, each person alive today is seen as quite distinguishable from everyone else. The degree to which this self-evidence about bodies' discreteness is engrained in much of our thinking is perhaps most readily revealed through how we often think about conjoined twins ("Siamese twins") (Box 2.3). Gould (1987: 200) used the example of a set of conjoined twins ("Ritta-Christina") who were born in the early nineteenth century to illustrate the problems of trying to conceptualize bodies as discrete entities. Specifically, he outlined how such twins' existence challenges ideas concerning bodies' individuality. Thus, should such twins be considered two people or a single entity? Certainly, conjoined twins have often been able to, for instance, speak and think quite independently, even to the point of arguing amongst themselves. Yet, when one twin dies the other inevitably dies soon thereafter, given how interconnected is their circulatory system. Whereas some commentators have sought to portray such twins as a single individual and others have portrayed them as two persons, Gould argued that such "categories were wrong or limited[, as t]he boundaries between oneness and twoness are human impositions, not nature's taxonomy." Hence, Gould averred,

> Ritta-Christina, formed from a single egg that failed to divide completely in twinning, born with two heads and two brains but only one lower half, was in part one, and in part two – not a blend, not one-and-a-half, but an object embodying the essential definitions of both oneness and twoness, depending upon the question asked or the perspective assumed.

Certainly, in either case, the twins' skin served as a boundary between them and the outside world, but this fact does not minimize the difficulties that such humans pose for thinking about the body as a scale – did their skin enclose and so separate from the outside world two individuals, one individual, neither, or both? The fact that the answer to this enquiry depends upon "the question asked or the perspective assumed" goes some way towards destabilizing the body as a natural spatial scale – changing the question asked or perspective assumed does not materially change the body of conjoined twins such as Ritta-Christina but it does dramatically change how we think of her/them, a reality that then shapes what claims we can make about bodies being discrete and self-evident bases for any hierarchy of spatial scales.

A second issue to consider with regard to the body as a coherent scale is that of the emergent properties of its constituent parts. Although bodies are typically considered single entities for purposes of being seen as geographical scales, they are also clearly made up of myriad separate bits and pieces – a brain, a heart, lungs, bones, muscle, blood cells, and so forth. Some of these can exist and "live," at least for a certain period of time, separate from the body from which they came – red blood cells, for instance, can be taken from one body and stored for several weeks before being used in a transfusion and placed in a second. Moreover, most of these various bits and pieces operate without our knowledge or control yet have a certain degree of what we might call, for want of a better word, "agency" – white blood cells, for instance, are programmed to seek out bacteria and kill them. To put this another way, upon considering how the body is constituted, we might think of it as an entity/scale which only exists as a result of the emergent properties created through the interplay of such constituent elements and because various of its constituent elements function independently of the consciousness of each individual human being. Furthermore, although the body's skin is reified as its container and delineator, its interface with the outside world, in fact the skin is not an impermeable outer boundary and is constantly being traversed by chemicals in both directions – some chemicals (sometimes highly toxic) can be absorbed through the skin whilst others (like water and salt) can be lost to the outside world through, for instance, sweating.

Taking this a step further, scientific evidence suggests that the mitochondria found within most eukaryotic cells developed from early proteobacteria that were subsequently absorbed into various human cells and evolved over billions of years as endosymbionts – the human body contains within its cells, in other words, DNA that originated in an

---

Box 2.3 CONJOINED TWINS: ONE BODY OR TWO?

Conjoined twins raise interesting questions about the nature of bodies – are they a single body or two bodies? Such twins also raise questions about representation – grammatically, should the singular "is" or the plural "are" be used? Perhaps two of the most famous examples of such twins are those of Chang and Eng Bunker (May 11, 1811–January 17, 1874), who were born in Thailand (then Siam), and Ritta-Christina Parodi, who were born in Sardinia on March 3, 1829 and died on November 23, 1829 (see Figures 2.2 and 2.3). For their part, Chang and Eng came to the United States and settled in North Carolina, where they took the name "Bunker." They worked with P.T. Barnum as one of his "curiosities." In 1843 they married two sisters, Adelaide and Sarah Anne Yates, eventually having twenty-one children between them. They invariably signed legal documents as "Chang Eng" (rather than as "Chang and Eng") and seem to have been somewhat ambivalent about the degree of autonomy they enjoyed. Ritta-Christina, meanwhile, were exhibited by their parents for commercial reasons and died on a trip to Paris. Because they had separate stomachs they were often hungry at different times. Although they were distinct from the shoulders up, they shared only a single set of genitals, a single anus, and one pelvis and set of legs. Both Chang and Eng and Ritta-Christina were said to have had quite different personalities. The question of whether such twins are one or two individuals led the Catholic Church to ponder whether Ritta-Christina had two souls or one (Gould 1987: 69).

---

entirely different life form and it has consequently evolved as a result of its symbiosis with such life forms (Margulis 1970). Drawing all of this together, these biological realities mean two things: i) the scale of the body itself only has functional coherence as a result of the coming together of myriad smaller- and larger-scale processes, for the body can only exist biologically as a result of what goes on within it and beyond it (the body is reliant on plants' photosynthesis, for instance, to turn the carbon dioxide it exudes into oxygen for its consumption); and ii) the scale of the body as marked by its external features is itself not as discrete as is frequently portrayed in the social science and humanities literatures

Figure 2.2
Eng (l) and
Chang (r) Bunker,
about 1860

Figure 2.3 (below)
Ritta-Christina,
from an illustration
in George M.
Gould and Walter
L. Pyle (1896),
*Anomalies and
Curiosities of
Medicine* (p. 185)

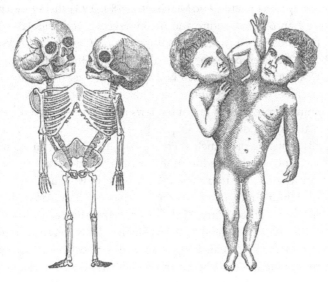

but is, instead, frequently crossed by chemical compounds, such that it is more difficult than perhaps first imagined to make the argument that the body is a distinct entity and that its covering of skin delineates where one space (the body's internal elements) ends and another (the outside world) begins. These facts raise questions of scale's gestalt – whether or not the body appears as a discrete entity depends somewhat on how it is viewed.

Continuing this question of how we might view bodies – and, so, the body as a scale – it is important to trace how the body has increasingly been conceptualized since the Renaissance (at least in the Western philosophical tradition) as an individualized entity separated from the world around it, for such a view encourages us to think about bodies – and, by extension, scales – as discrete things. In particular, Merchant (1980) has argued that a fundamental rethinking of the nature of Nature and of matter took place in seventeenth-century Europe. Specifically, the Renaissance saw the emergence of what she calls a "mechanical order," one which radically transformed how people viewed their relationship with Nature and in which the human body increasingly came to be conceptualized in terms of a new metaphor – that of "the machine." Supported and diffused by the writings of philosophers such as René Descartes, Marin Mersenne, and Pierre Gassendi, the mechanical order supposed that the world was not an organic whole, made up of particles which were alive (a common view in the Middle Ages), but, rather, was constituted by inert matter. For Merchant (1980: 193) this represented the death of an organic Nature and the emergence of a popular *weltan-schauung* ("world view") in which "nature was now viewed as a system of dead, inert particles moved by external, rather than inherent [,] forces." In turn, this resulted in the human body becoming ever more conceived of as a machine whose operations were the result not of some cosmic spirit running through them but of the laws of an external Nature working upon them – bodies were shaped by external and universal laws. Within this context writers such as Francis Bacon argued for an atomistic view of social objects divorced from the broader world of Nature, a view which marked a distinct break from the holistic view of human bodies as part of a more integrated cosmos which had dominated during the Middle Ages. Consequently, "[a] new concept of the self as a rational master of the passions housed in a machinelike body began to replace the concept of the self as an integral part of a close-knit harmony of organic parts united to the cosmos and society" (Merchant 1980: 214).

For sure, bodies were subject to the laws of physics in the same way in which all other elements in the universe were, and in that sense could be seen to be part of a material continuum. However, they were no longer organically and integrally connected to the broader cosmos/society but were, instead, seen as individuated organisms subject to that cosmos's/society's laws and forces. The body, in other words, had been conceptually separated from the context within which it found itself.

Other changes took place in Western Europe with regard to how the body was conceptualized. Most particularly, there was a transformation in how male and female bodies were considered, a change from what Laqueur (1990) has called a "one-sex" to a "two-sex" model. Specifically, Laqueur maintains, beginning with the ancient Greeks, women's bodies were considered to be flawed varieties of male bodies within a hierarchical binary. For instance, whereas male and female reproductive organs were seen as homologous, those of the female were considered inferior because they had not emerged outside the body – women, in other words, were simply less advanced/developed versions of men. Likewise, other parts of the body were seen to have a correspondence in males and females but to play radically different roles in each – blood was viewed as the fluid from which semen arose before it passed through the brain, spinal cord, and kidneys into the testes, whereas breast milk was understood to be womb blood that flowed to the breasts, where it became white. However, around 1800 medical and other writers increasingly began to insist that there were fundamental differences between men and women that were based upon biological distinctions – men's supposed passion and quickness and women's passivity and slowness were now reinterpreted in terms of a radical dimorphism based upon biology. The result was that whereas male and female organs such as the penis and vagina had previously generally been viewed as parallels or mirror images of each other, now they increasingly came to be seen as opposites.

It is important to recognize that such views of the body as not connected organically and integrally to the broader landscape do not predominate in all cultures, despite the history of intellectual colonization to which they have often been subjected by Western Europeans. For instance, the Zoque-speaking Indians of Chimalapas in southern Mexico envision themselves as ontologically connected to the land, not as separate from it. Thus, they suggest, "We are Chimalapas and Chimalapas is us – our ancestors are in this land as are we, with the forest, the animals, the trees" (quoted in Walker and Walker 2008: 167–68). Consequently,

for Zoques "there is no distinction between space, place and being."
In such a worldview, "Nature cannot be bounded and separated from
the body." Perhaps more importantly for our purposes here, however,
such a worldview means also that the body cannot be bounded and
separated from Nature and the world around it. This raises significant
implications for conceptualizing the body as a geographical scale, for it
suggests that views of the body as a discrete ontological unit are highly
culturally specific. Likewise, in many African societies it is believed that
individuals are never really so individual because they are connected
supernaturally with the spirits of their ancestors, such that there is an
inexorable bond between the two and neither exists separate from the
other. Equally, the Zulu concept of "Ubuntu," in which each human is
seen as inextricably connected to all others – as in the Zulu maxim
"umuntu ngumuntu ngabantu" ("a person is a person through other
persons") – directly contradicts Western Cartesian notions of the
individualized body. Specifically, whilst ubuntu's,

> respect for the particularity of the other links up closely to its respect
> for *individuality*[,] . . . the individuality that *ubuntu* respects is not of
> Cartesian making . . . [Thus, whereas t]he Cartesian individual exists
> prior to, or separately and independently from[,] the rest of the com-
> munity or society[,] . . . *ubuntu* directly contradicts [this] conception
> of individuality in terms of which the individual or self can be
> conceived without thereby necessarily conceiving the other.
>
> (Louw 2006: 168)

Moreover, even within "the Western tradition" the idea of the body
as an atomistic entity is not hegemonic. Hence, we can rhetorically ask,
how might we think of bodies not as atomized and individualized
biological entities but as ensembles of biological material which has come
from other bodies? Given that we get our DNA from our biological
parents, who got theirs from theirs, who in turn got theirs from
theirs, what does this say about us as discrete entities? What does the
fact that our own bodies are constituted by the DNA acquired from two
other bodies mean for how we think of our own place in relation to our
ancestors? Certainly, even if the DNA acquired from parents combines
in unique ways in each child and, in that sense, we might all of us be
seen to be individual biological entities, we nevertheless perhaps occupy
a sort of shadowland presence – we can be viewed as unique

Figure 2.4 The temporal connection of humans across the generations, as represented in an image of evolution (though note that, in such diagrams, a male figure is typically taken to represent all of humanity and evolution is represented in terms of a single evolutionary line rather than the [now] more accepted tree of human evolution)

combinations of genetic material, but that genetic material itself is not unique to us, for it has been passed down through the generations. As a result, although we may well be separate individuals, we are also part of our parents, as they are part of us. In following our own genetic trail backwards into history, we discover that, to quote Sierra Club founder John Muir (1916: 157), "[w]hen we try to pick out anything by itself [such as our own body], we find it hitched to everything else in the universe." Our genetic umbilical cord links us inalienably to all the past generations who have gone before us, even as our corporeal umbilical cord may have been cut decades ago. It is perhaps not surprising, therefore, that a popular visual image to explain human evolution often represents a line of pre-modern humans standing behind a modern human (see Figure 2.4).

Taking this a step further, however, even a cursory knowledge of biology indicates that we are each of us dependent upon things beyond our individual bodies for life for, as Marx (1844/2007: 74) suggested, the human body has to "remain in continuous [contact]" with Nature "if [it] is not to die." There are, in other words, necessary linkages between us and the world beyond us – we need food, water, and sunlight to survive. This suggests that how we conceive of the body depends upon how we abstract it from the broader set of biological and social relations within which bodies exist materially and how we choose to view it:

as an individual entity or as part of a continuum – dare we say network? – of life. What is more is that not only are bodies as biological entities connected to other biological entities – their parents, food sources, and so forth – but, increasingly, human bodies are being connected to manufactured entities, not just in terms of their using such entities but in terms of them actually being incorporated into bodies, with the result that the division between the biological and the manufactured is being ever more transgressed. Such technological and social developments take us into the world of what Haraway (1991: 149) and others (e.g., Hughes 2004) have called the cyborg, a "cybernetic organism, a hybrid of machine and organism." Although Haraway used the idea of the cyborg principally as a mechanism by which to destabilize various common binaries (object/

Box 2.4 THE BODY AS CYBORG

The Canadian media commentator Marshall McLuhan (1964/2001: 3) once commented that humans "have extended our central nervous system itself in a global embrace, abolishing both space and time as far as our planet is concerned." McLuhan was talking about how technologies such as the telegraph and the telephone have allowed humans to cross time and space. However, developments in cybernetics and bio-implants represent a fundamentally different stage in this technological advance, as they quite literally transgress the boundary of the body's skin which previously separated humans from technologies such as telegraph machines. In 1998, Professor Kevin Warwick of the University of Reading (UK) installed an RFID (Radio-frequency identification) chip into his forearm which allowed him automatically to open doors, manipulate lighting, and adjust the temperature in a room simply by entering it. He termed this experiment "Project Cyborg 1.0." He subsequently engaged in "Project Cyborg 2.0," in which a micro electrode array was implanted in the median nerve of his left arm. This enabled him to send signals directly from his nervous system to various electronic objects and even to other human beings with implants connected to their nervous systems. Such developments raise questions about whether the body as a scale is an areal unit, with its external boundary enclosing an internal space, a node/location on a broader network, or both.

subject, Nature/culture) in ways that she felt were useful for feminists, the history of medicine shows that human beings have more and more become cyborgs over time – we inject our bodies with various manufactured vaccines and antibiotics to avoid becoming sick or after we have become so, we replace defective knees, hips, and hearts with mechanical ones, and we implant cochlear devices when we are profoundly hard of hearing. Already, there are efforts being made to grow – that is to say manufacture – new organs (livers, kidneys, etc.) in the laboratory to replace malfunctioning ones and to develop transponder implants for humans so that they may take advantage of emerging mobile commerce applications (Michael and Masters 2005). Indeed, some humans have already been implanted with chips that can send radio signals to computer-controlled devices to operate them (Warwick 2004), whilst chips containing GPS technologies that provide some security against children or business people being kidnapped and/or medical information that might proactively communicate with computers – for instance, to indicate changes in blood chemistry or body temperature – are also being developed and marketed (Box 2.4). Such developments raise to an even higher plane questions about where an individual body ends when it can be in constant contact with, quite literally, anyone in the world who is hooked into a computer network and may remotely activate various electronic devices, and what this means for how the body as a scale is conceived of.

CONCLUDING REMARKS

The above discussion raises several important issues with regard to thinking about the body as a geographical scale within a collectivity of scales.

First, there is clearly a divergence in the literature between those who focus upon the body's biological elements and see bodies as natural building blocks upon which cultural elements may be superimposed – the "body as cultural coatrack" approach – and those who see the body's biology as socially produced. This "body as natural unit" versus "body as socially produced" dualism mirrors debates over whether scales are natural entities or whether they are socially produced. Furthermore, it raises questions about how certain aspects of the body as a body are reified and what that means for thinking about the body as a coherent scale. Hence, the body as a biological entity is often seen to have a certain gestalt – despite the fact that it is made up of trillions of individual cells, when viewed as a totality it is considered to have a degree of wholeness and physical integrity. Put another way, the body is only understandable as a body from a certain (external) perspective, for only if we think of its quite different constituent parts as working in harmony beneath the skin to allow the greater whole to function does it appear to exhibit a oneness. Such a reification of the body as a unit is in stark contra-distinction to a viewpoint that might focus upon various components of the body (the heart, the lungs, the eyes, etc.), for from this subepidermal viewpoint the body's integrity appears to be undermined – it is little more than a collection of disparate parts, each with quite different functions. This view is analogous to Cosgrove's (1994) distinction within the globalization literature between the concepts of "one-worldism" and "whole-Earthism," with the former seeing the global as much more func-tionally integrated than the latter (see Chapter 6). Equally, in the case of scales, it is often argued that what gives scales their coherence is the fact that they suggest that what is within them has a certain degree of simi-larity compared to what lies beyond them – for instance, regions are often seen to enclose areas that share greater similarities than do areas beyond them. The reification of the body's oneness and that of scales' oneness, then, comes from focusing upon their exteriors rather than interiors.

Second, and following the first point, much of the literature about bodies has presented understandings which put forth significant dualisms. For example, within the psychoanalytical literature there has been a long-standing binary between the conscious and the unconscious, whilst more generally there has been a binary between the body and the mind. Other binaries have also been evident, such as male/female (a binary which ignores other genders), white/non-white (a binary which privil-eges whiteness), normal/abnormal, abled/disabled, and straight/queer, all of which have implications for, and parallels with, thinking about the body as a scale – such as when the world is seen in terms of a local/global binary (I shall return to this in Chapter 6). One of the most noteworthy binaries has been that of inside/outside, a binary which sees the skin as the delineator between the body's inside and the outside world, as the container of the body's internal spaces. Tellingly, this inside/outside dualism and viewing the skin as the body's container parallels debates about scales more generally, especially concerning whether the body qua scale is viewed in areal/container terms as being a scale at the lower end of a series of other scales conceived of in areal/container terms, or whether it is seen as a node or actor within a broader network of biological and social relations à la actor-network theory (Latour 1996). Thus, the views explored by Merchant (1980) concerning how the body increasingly came to be seen as no longer connected organically to everything else through some kind of universal spirit that flows through it but, rather, as a discrete entity subject to external forces or processes perhaps fits more closely with conceptions of scales as containers – bodies are, in such a perspective, little more than containers of matter – than it does with conceptions of scales as part of a broader network. On the other hand, given that each individual body is made up of genetic material which has been passed down to it by its parents, who themselves received genetic material from theirs, then each body may be seen to be connected into a network of humans going back across time. Similarly, given that we also share genetic material with the other descendents of our long-dead ancestors – a single ancestor born three or four hundred years ago may have several thousand direct descendants alive today – we are also connected into a network of family members across (often vast) expanses of space in the contemporary period. Equally, perspectives such as that of Haraway, who views bodies as cyborgs, also seem to be more consonant with Latourian views of scale as networked than with those which view scales as relatively fixed and discrete containers of matter/space. By the same token, French philosopher Merleau-Ponty's (1945/2002: 235) view that each human is part of a single system in which they are connected to others through the nature of their being on Earth's "common ground" and in which each body is produced by the inter-actions with others within the world bears similarities with Latour's conceptualization of scales as fibrous, thread-like, and capillary networks. Questions about whether the body is seen to be a discrete ontologic entity and container (in other words, can it exist independently of other entities?) or whether its interconnectivity to the biological and social world around it means that it cannot be so viewed, have their parallels in discussions about whether particular scales may be seen to be ontologically discrete or not.

If the biology/social dualism is one important binary wherein there is some overlap between thinking about bodies qua bodies and thinking about bodies qua scales, and another is the internal/external dualism (which leads to issues of whether the body's skin should be viewed as serving as the boundary layer for/container of the body and, depending upon the answer to that question, whether the body itself should be viewed in terms of a container or in terms of being a node in a network), then an additional important dualism is that concerning gender. Specifically, one way in which the body has been considered as a body is through its gendering. Significantly, though, scales have also sometimes been viewed through the lens of gender. In particular, the association of strength and power with the global and, in contrast, of weakness and impotence with the local (Gibson-Graham 2002) have often resulted in the global being associated with the masculine and the local with the feminine, a discourse which is both gendered and heteronormative (I shall return to this in Chapter 6). There seems, then, to be a degree of cross-fertilization regarding how we might view bodies as bodies and how we might view the body as a scale, given that the body is frequently seen to be either a stand-in for the local or the smallest scale in a scalar hierarchy that links it with its Other, the global (as with Harvey's (2000) formulation – see below). Moreover, and to push the argument a little further, given that the glocal scale is often conceived of as containing elements both of the global and of the local, we perhaps might even consider it as something of a hermaphroditic scale.

Third, there is also a degree of conceptual cross-fertilization between considerations of the body as dialectically constituted and constituting – that is to say, the body as an entity which is both shaped by, but also shapes, Nature – and scales as being dialectically constituted and constituting. For example, Harvey (2000) has argued – as did Marx – that bodies under capitalism both serve as the origin of value and are also disciplined and shaped by the operations of global capitalist social relations. Furthermore, processes such as globalization (Herod 2009) may be initiated by individuals but they are about reworking at a planetary scale the social and biological relationships between billions of individual bodies. As Harvey (2000: 109) puts it: "[s]patiotemporality defined at one scale (that of 'globalization' and all its associated meanings) intersects with bodies that function at a much more localized scale" – the relations of globally organized capitalism, in other words, shape the life possibilities of individual bodies even as such bodies shape how globally organized capitalism operates. At the same time, however, whilst there is a consonance between dialectical approaches to the body and to scales (the body as body and as scale is seen to shape, but is also shaped by, other scales), there is also a consonance between non-dialectical approaches to both – whilst much theorizing about the body has seen the biological and the social/cultural as two separate realms, much scalar theorizing has seen various scales such as the urban and the global as separate realms (à la the "scale as ladder rungs" or "scale as concentric circles" approaches) rather than as dialectically and inexorably connected wherein change in one scale reflects or heralds change at another.

A fourth point of contact between considerations of the body qua body and the body qua scale is the issue of the body and scales as tropes. Thus, whereas Jones (1998: 27) suggested that "scale itself is a representational trope, a way of framing political-spatiality that in turn has material effects" and shapes meaning and understanding of particular situations and processes, the body has likewise served a similar role. Hence, the body has often been used as a trope for the economy – as when the circulation of blood has been used as a metaphor for the circulation of capital throughout the economy or when the economy has been described in biological terms. Indeed, in this particular case, Amariglio (1988: 585) has argued that "modern economic theory in most if not all of its variants (neoclassical, Marxian, Austrian, Institutionalist, Post-Keynesian, and so forth) is grounded partly in a concept of the body" and has suggested that it is possible even to "differentiate alternative positions in economic theory since [Adam] Smith and [David] Ricardo in terms of the type of body and its internal relations." Significantly, however, one of the most common tropes involving the body – at least as it relates to scale – is that of the body as a trope for the nation/nation-state (I shall return to this in Chapter 5). For instance, the medieval philosopher John of Salisbury (1159/1990: 66–67) maintained that "a republic is . . . a sort of body" in which the head of state can be thought of as being like the head in a human body, whereas the senate represents its heart, the judges and governors of provinces its ears, eyes, and mouth, bureaucrats and soldiers its hands, treasurers and record keepers its stomach and intestines, and the peasants, who are "perpetually bound to the soil," its feet.

Likewise, as Cañeque (2004: 21) has argued, during the sixteenth and seventeenth centuries the Spanish Crown used the metaphor of the body to shape political institutions both at home and in the Americas, with political power being dispersed to various entities but being focused through a single head – the monarch. Hence, the notion of the *corpus iuridicum* (body politic) prevalent in Spanish political thought at this time argued that individuals only exist in society as "members of a body and that the hierarchical organization of the political community is as natural and well ordered as that of a human body, which, in turn, is a reflection of the perfect ordering and harmony of the celestial bodies."

Given that such a "mystical body would . . . be incomplete without a head, the king[,] . . . this organic unity of head and limbs in the political community[, could] always [be] used as the main argument to justify the advantages of monarchical rule." The dispersion of political power to various relatively autonomous centers which were coordinated through the body politic's head (the monarch) "corresponded to the dispersion and relative autonomy of the vital functions and organs of the human body, which . . . served as the model for social and political organization" (p. 76). Significantly, for our purposes here, such tropes not only undoubtedly shaped how people understood their own bodily relationship to the state and so had material consequences, as Jones (1998) has suggested is the case with scalar representations, but they also represented one scale, that of the national, in terms of another, that of the body.

Other parallels between how the body has been conceptualized and how scales have been conceptualized can be seen with regards to the erasure of both within a supposedly globalizing world. Hence, Virilio has suggested that the speed which characterizes contemporary processes of globalization is leading to the death of the body. Likewise, the fetishization of the commodity spoken of by, amongst others, Marx that has intensified as globalization has stretched production chains across the planet (so making it more difficult for consumers to know much about the conditions under which the commodities they consume are produced) has also made the body increasingly invisible – thereby metaphorically killing it – in the contemporary economy. This corresponds with arguments that globalization is bringing about the "delocalization" (Virilio 1997; Gray 1998) and/or "denationalization" (Sassen 2003) of economic and political life, with the result that rather than history taking place "within local times, local frames, regions and nations," from now on it "is going to unfold within a one-time-system: global time" (Virilio 1995b). Many discourses of the global erase both bodies qua bodies and bodies qua scales. In this vein, we might also consider how discourses of "the body" rather than of "bodies" render individuals' bodies invisible by assuming there to be a kind of normative ideal-type body that exists. Thus, the concept of "the average man/woman" serves to cloak real people's bodily characteristics behind a representative fiction – to take a simple example, in a room in which three people are 6 feet tall and three are 5 feet tall, the average height is 5'6", even though not a single person in the room is actually this height.

Finally, a sixth point of convergence between theorizations of the body qua body and of the body qua scale relates to how both bodies and scales have been conceptualized in terms of their performativity. Thus, whereas Butler (1990) has argued for viewing the body's gender through the lens of performativity, Kaiser and Nikiforova (2008) have likewise advocated viewing scales as a category of practice through which scale effects are socially produced – it is through the repetition of sets of discourses concerning scale, they aver, that scale effects are produced and deployed as devices that shape place and identity. Much as bodies may perform their gender, so too can scales be seen to perform by shaping how social actors understand their relationship to others.

3

THE URBAN

Cities are the central elements in spatial organization of regional, national, and supranational socioeconomies by virtue of the interregional organization in a total "ecological field" of the functions they perform.

Berry (1965: 111)

Although the industrial city was a centrepiece of accumulation and surplus production, it has to be seen as a distinctive place within the spaces of the international division of labor.

Harvey (1989a: 30)

Cities have long been of central interest not just in Geography but also in a host of other fields. Consequently, the urban scale has been a focus of intense theorizing, regarding both how to conceptualize cities themselves and how to conceptualize the urban as a scale within a broader panoply of scales. Much of this theorizing has involved contemplating whether urban areas are distinct and separate realms and whether they play specific roles within particular societies – as in questions of whether they form one end of, say, a rural–urban continuum of spaces, such that rural and urban spaces are viewed somewhat as mirror images/polar opposites of each other through which to interpret society (e.g., Tönnies 1887/2001; Redfield 1941). Given this, in this chapter I begin with a brief view of how urban areas have variously been conceived of historically. I then explore how the urban has been theorized in the twentieth century, paying particular attention to the Chicago School of social ecology, and Marxist and other critical approaches. Finally, I examine the growing use of the network metaphor to understand urban dynamics. The chapter ends with a brief consideration of what this all means for theorizing "the urban" as a scale.

EARLY VIEWS OF THE URBAN

It is generally accepted that cities were first established in the Bronze Age in the Middle East. However, throughout history they have been conceived of in quite different ways. In the Bible, for instance, cities appear early – the Book of Genesis mentions that Cain, after murdering his brother Abel, founded the world's first city – and four different urban archetypes were used to tell various stories about sin, morality, and redemption: Sodom, a corrupt city without righteousness, which deserved divine punishment; Babylon, a cosmopolitan city in which anything and everything could be found; Nineveh, the metropolis which repented and so was welcomed back into the community of the virtuous; and Jerusalem, a city founded on righteousness and designed to be a spiritual center. As Elazar (2004) notes, throughout the Bible there is "a paradigm of urbanism in relation to biblical morality." This morality continued iconographically in Europe at least until the medieval period, with maps of the world frequently placing Jerusalem at its center, drawing upon the Gospel of Luke's presentation of the city as a kind of *axis mundi* (world axis) (see Herod 2009: 32 for an example). Meanwhile, in other parts of the New Testament, Jerusalem is seen as the central point from which Christianity will diffuse across the world (Acts), whilst a New Jerusalem will be created at the end of the world (Book of Revelation).

For their part, although their various city-states were governed quite differently, the ancient Greeks generally saw the city (the *polis*) as the physical expression of a community of citizens and thus as the hearth of culture, although they had a somewhat flexible concept of what a city might be in terms of its spatial extent – the *polis* could stretch in size from a settlement with a few houses to a large urban center. In Aristotle's eyes, the *polis* was like "an organism, coming into life [in] elementary form, but already with the seeds of its future growth[,] . . . a composite whole made up of various parts in varying relationships" (Vlassopoulos 2007: 78) – a representation that brings to the fore some of the issues about the body discussed in Chapter 2. Furthermore, because the *polis*, as the embodiment of the community, was seen to have territorial expression, it was also viewed in areal terms as serving as a spatial container of particular communities – a Greek and a Persian, for instance, could be friends (*xenoi*) but not members of the same *polis*. In terms of the city's physical layout, at its center would typically lie the *agora*, a physical space – sometimes circular, sometimes rectangular or triangular – for commerce and discussing important issues. The *agora* itself was based on the older idea of the warriors' assembly, in which a warrior would stand at the center of a circle and speak freely to his fellow warriors. However, with the *agora*'s advent, initially in the Greek colonies but later in the mainland cities, a significant transformation of urban space in Greek cities and of concepts of the city took place, such that the Greeks increasingly visualized the city as an areal representation of the cosmos in which stress was put on the center and the geometry of the circle, with the *agora* metaphorically sitting as the hub of the world out from which, in concentric fashion, the rest of the landscape flowed in geometric progression (Naddaf 2005: 83). This was a very different view of space than that found in some of the older Asian monarchies, in which "the structure of the terrestrial states reflect[ed] the celestial state," with the universe "perceived as a hierarchy of powers analogous in its structure to a human society." The result was "not a universe with a homogeneous space, but a universe [viewed] in the form of a pyramid, or different levels." Given the homologous nature of the terrestrial and celestial worlds in such views, the end result was "a close solidarity between the physical and political spaces," such that "they both reflect[ed] a pyramidal structure." As with viewing the *polis* in biological terms, so too here were some very clear scalar understandings of the world incorporated in urban areas' physical layout – a Greek concentric circle view of the world versus an Asian ladder-like one. As Naddaf (2005: 84) puts it:

> The *agora* is . . . the . . . centered space that permitted all citizens to affirm themselves as *isoi* (equals), and *homoioi* (peers), and to enter with one another into a relation of identity, symmetry, and reciprocity [as part of] a united cosmos . . . The *agora* is thus the symbol of a spatial structure radically different from the one that characterizes the oriental monarchies. The power (*kratos*, *archē*, and *dunasteia*) is no longer situated at the top of the ladder. The power is disposed *es meson*, in the center, in the middle of the human group.

The Romans likewise saw the urban as playing a distinctive social role – the Latin word *civitas* is, after all, the origin of the English word "civilization" – and had three quite different ways of describing various urban areas (in rank-order, from most to least important: *colonia*, *municipium*, and *civitas*). Roman city planners used grid systems and building codes to impose a visual and functional order on the built environment, with such urban order being a metaphor for the order they wished to impose throughout their vast empire. Hence, whereas the Greeks and Etruscans viewed cities as more or less independent entities within broader federations, the Romans had a much more highly developed sense of spatial hierarchy, with each urban area playing a particular role within the empire and the city of Rome itself serving as the imperial focal point. Indeed, Romanness (*Romanitas*) was expressed explicitly through connections to Rome (as in the phrase *Civis Romanus sum* – "I am a citizen of Rome"), with citizens throughout the empire having the right, for instance, to be tried in Rome. More generally, living in any city was considered crucial to one's ability to fully articulate a sense of being civilized, and the metaphor of the city was often used to express broader philosophical concepts – hence the emperor and stoic philosopher Marcus Aurelius, in an interesting scalar shift, suggested that "[b]ecause the universe is like a city," all of the cities across the world were like "the houses of one city" (quoted in Thomas 2007: 147).

Beyond Europe, the Mayan cities of the Classic period (250–900 AD) existed rather like Greek city-states, with residents unified by common cultural characteristics (such as language) but generally retaining separate identities. The Maya often viewed cities as places from which royal power emanated, with cities such as Calakmul and Copán designed such that their centers were used chiefly as a space for religious and civic ceremony, with most of the population living outside this center and in villages which surrounded it. The Conquistadors, meanwhile, saw cities as their link to civilization, and actively promoted fairly strict urban planning – colonial Spanish cities typically were based on a grid, with a large plaza at the center serving as a market and also as a place of execution, thereby reinforcing symbolically the power of the Crown over individual subjects (Burkholder and Johnson 2003: 236). In the process, they often built their own cities on top of those of the indigenous population, destroying vast sections of them even as some vestigial elements of the pre-colonial cities were incorporated into the Spanish ones.

Clearly, urban areas have been important elements within human history and have represented different things in different cultures at different times. In terms of contemporary social theory and geographic thought, though much theorizing about cities has been concerned with their economic roles under capitalism, given that the defining feature of capitalist development has, arguably, been its urbanization. For instance, in *The German Ideology* Marx and Engels argued that the growth of towns beginning in the late Middle Ages reflected a crucial development in the division of labor that would ultimately lead to capitalism's appearance. Hence, whereas towns had clearly existed prior to capitalism's emergence, they had not been the focal point of the extant mode of production in quite the way in which they became under capitalism. As Marx and Engels (1845/1970: 43) put it:

> The division of labour inside a nation leads at first to the separation of industrial and commercial from agricultural labour, and hence to the separation of *town* and *country* and to the conflict of their interests. Its further development leads to the separation of commercial from industrial labour. At the same time through the division of labour inside these various branches there develop various divisions among the individuals co-operating in definite kinds of labour.

For his part, Durkheim (1893/1947) focused on how growth in towns transformed long-standing social bonds between people whose ancestors had lived in the same communities for centuries and replaced them with feelings of *anomie* in which social expectations were increasingly unclear because of the mixing within cities of people who did not know each other. Weber (1921/1958) was likewise interested in the historical development of cities, arguing that they were central spaces in breaking the grip held on society by medieval political and economic institutions and in allowing a new rationality to develop in Western Europe, developments which proved central to the emergence of capitalism and liberal democracy. The city was, he suggested (p. 1220), "a partially autonomous organization," somewhat separate from the broader society within which it sits. Based upon this, he developed an urban typology, distinguishing between the "consumer city" and the "producer city," the "industrial city" and the "merchant city."

Although in many ways their approaches were quite different, Marx, Durkheim, and Weber did share similarities when theorizing the urban.

Thus, all three saw the city as a historically important object of analysis in explaining the transition from feudalism to capitalism in Western Europe. Equally, all three saw rapid and widespread urbanization not as a cause of the emergence of capitalism but as a consequence thereof – hence, for Marx, the growth of cities did not create the modern proletariat but did allow it to understand its own position in society, especially in contrast to that of capital (Saunders 1986). At the same time, their approaches were all rather Eurocentric, since they based their understandings of society, capitalism, and urbanization's role therein upon the role cities had played in Europe (Giddens 1981). Thus, as Käsler (1988: 43) has pointed out, "Weber worked from the assumption that every city – in Antiquity as in the Middle Ages, inside and outside Europe – was a kind of 'fortress' and 'garrison', i.e. a castle or wall belonged to the city and Weber regarded 'the seigneurial castle', and with it castle-seated princes, as a universal phenomenon." Likewise, Durkheim's dualism of "mechanical solidarity" (wherein all members of a society perform more or less the same tasks, such that society's collective conscience is virtually identical amongst all members) and "organic solidarity" (wherein society is extremely divided in terms of social responsibility, such that individuals no longer perform the same tasks nor necessarily share the same perspectives on life) reflects his understanding of the development of the division of labor within Western Europe. Such Eurocentrism raises important questions about the role played by the urban and, consequently, efforts to theorize this scale critically within the broader pantheon of scales.

THE URBAN AS A FOCUS OF STUDY IN THE TWENTIETH CENTURY

Given such theoretical interest in cities' role under capitalism, together with the fact that much of the Western world experienced rapid urbanization in the nineteenth century, by 1900 the urban had become an important analytical object. For example, the urban as a scale of analysis was quickly formalized through the creation of courses focusing upon cities – the first university programs devoted to studying cities in the Anglophonic world were created in 1909, when Liverpool University established a degree program in town and country planning and the Massachusetts Institute of Technology instituted a course in urban planning, followed by the University of Chicago's establishment of a center for studying urban sociology in the 1920s (LeGates 2003). In this light, one of the earliest academic efforts to think theoretically about urban spaces and how they are different from other types of spaces – that is to say, how we might think of cities as absolute spaces that are distinguishable from the spaces surrounding them – was that of Wirth (1938), who opined on "urbanism as a way of life." Hence, he (p. 3) suggested that the city and the country could "be regarded as two poles in reference to one or the other of which all human settlements tend to arrange themselves." More significantly, he (p. 4) argued against defining something as urban on the basis of size alone, which he saw as "obviously arbitrary," suggesting that "[a]s long as we identify urbanism with the physical entity of the city, viewing it merely as rigidly delimited in space, and proceed as if urban attributes abruptly ceased to be manifested beyond an arbitrary boundary line, we are not likely to arrive at any adequate conception of urbanism as a mode of life." Instead, he sought to define the urban on the basis of process – "the forms of social action and organization that typically emerge in relatively permanent, compact settlements of large numbers of heterogeneous individuals" (p. 9). This is significant, for although Wirth did not use the Marxian language of the social production of scale, his effort to distinguish between the spatial extent of cities and the urban as a coherent scale is precisely what writers such as Smith (1984/1990) have tried to do (see below).

The Chicago School

The group of scholars at the University of Chicago who formed what became known as the "Chicago School" (of which Wirth was a part) provided some of the most sustained analyses of cities and the urban in the early twentieth century (Box 3.1). Drawing upon ideas from plant ecology and Darwinian evolutionary thought, they developed the idea of "human ecology" to understand how cities are structured and how they grow. This approach likened the city to a biological organism, suggesting that each of its constituent neighborhoods played a role in the city's functioning and was therefore symbiotically related to the city's other neighborhoods. Although city residents were, they suggested, driven by competitive urges to pursue their own ends as part of a "survival of the fittest" strategy, especially in the economic realm, they also were capable of cooperation, which generated a moral order that did not let competition become too extreme. Thus, unlike plants, urban dwellers would develop a cultural "superstructure" which would serve to moderate their behavior (Park et al. 1925). Drawing upon these theoretical ideas, and using Chicago as their laboratory, members of the School set about trying to understand how urban areas functioned. Specifically, they proposed that different neighborhoods become home to different ethnic groups through a process of "invasion" and "neighborhood succession," in much the same way that natural landscapes are transformed through the invasion of new plants and animals as they move toward their climax community (Clements 1916).

Based upon these ideas, School members developed an understanding of urban growth in Chicago which they believed applicable to other cities. Specifically, they furthered a model based upon concentric circles emanating from the urban center, with each circle representing a particular type of land use. Hence, at the city's center stood its Central Business District (CBD). This was surrounded by a "Zone in Transition," wherein industry and cheap housing were to be found. Beyond this were the "Zone of Workers' Homes," the "Zone of Better Residences" for middle-class families, and, finally, the "Zone of Commuters." As the city grew over time, the CBD would begin to invade nearby neighborhoods in the Zone in Transition immediately beyond it, which would, in turn, lead to a wave of development outwards – older middle-class neighborhoods might become downgraded and so a destination for immigrants as the middle class moved further out. In response to this model, Hoyt (1939) developed a "sector model" of the city, in which, rather than circles radiating out from the CBD, wedges of relatively homogeneous land use did so, as particular land-use types followed transportation routes out towards the suburbs. In many respects, however, Hoyt's model was little more than a modified concentric zone model, with the circles distorted by the presence of transportation corridors – Hoyt, for instance, retained the notion that the city develops according to ecological succession principles. Finally, Harris and Ullman (1945), both also trained at Chicago, developed the "multiple nuclei" model of urban development, which suggested that specialized zones of activity develop according to different activities' abilities to pay rent or their need to cluster together/desire to remain separated. Although the principal CBD remains at the city center, other "mini-CBDs" are seen to develop beyond it, often located in the urban periphery. They too retained the basic ideas of ecological succession (Carr 1997).

Box 3.1 THE CHICAGO SCHOOL OF HUMAN ECOLOGY

Beginning in the early twentieth century, the Department of Sociology at the University of Chicago – founded in 1892, the first such department in the US – began to focus upon urban sociology, combining urban theory with ethnographic fieldwork. This approach, drawing heavily upon ideas from biology, was known as Human Ecology. The School's early days are most commonly associated with the work of Robert Park, Ernest Burgess, Louis Wirth, and Homer Hoyt, though others were also involved.

Robert Ezra Park (1864–1944): educated at the University of Michigan, he became a journalist in Chicago and wrote about matters of race and the urban environment. He subsequently did graduate work at Harvard and in Germany, where he worked with the neo-Kantian urban sociologist Georg Simmel (a friend of Max Weber) and the geographer Alfred Hettner (best known for developing the concept of chorology (see Chapter 4)). He briefly taught at Harvard and then at the Tuskegee Institute in Alabama, before becoming a professor at the University of Chicago in 1914. He was president of the American Sociological Association (ASA).

Ernest Burgess (1886–1966): born in Canada but educated in Oklahoma and also at the University of Chicago, Burgess was hired as a professor at Chicago in 1916 and worked closely with Park. Also president of the ASA, Burgess is most associated with developing the Concentric Zone Model of urban growth, based upon Chicago's historical evolution. However, he was also interested in studying the institution of marriage and the effects of retirement on people.

Louis Wirth (1897–1952): originally from Germany, he came to the US in 1911. He principally focused upon how Jewish immigrants adjusted to life in urban America and advocated applying sociological theory to solve realworld problems. Wirth viewed urbanism as harmful to traditional culture and ways of life through, for instance, weakening kinship bonds and leading to a decline in the family's social significance, but he also viewed cities as being emblematic of what is distinctively modern in civilization.

Homer Hoyt (1895–1984): an economist who worked in real estate, Hoyt received undergraduate and graduate degrees from the University of Kansas before earning both a J.D. (1918) and a

continued

Box 3.1 continued

Ph.D. (1933) from the University of Chicago. In 1934 he joined the Federal Housing Administration as a housing economist. A decade later he became a visiting professor at the Massachusetts Institute of Technology and Columbia University. He is perhaps most well known for developing the sector model of urban land use, which, amongst other things, suggested that land values and a neighborhood's racial composition were closely linked. However, he also helped develop economic base theory and techniques for real estate market analysis.

Chauncy Harris (1914–2003): born in Utah, Harris was a Rhodes Scholar at Oxford University. He also studied at the London School of Economics and completed a Ph.D. at the University of Chicago in 1940. He taught at Indiana University and the University of Nebraska before being hired at the University of Chicago in 1943. He wrote much about the urban geography of the Soviet Union and was a president of the Association of American Geographers.

Edward Ullman (1912–76): Ullman is largely credited with bringing Walter Christaller's Central Place Theory to prominence in the United States. Although he was trained at Chicago and most famously devised a model of urban structure with Chauncy Harris, he spent his academic teaching career at Harvard (until the department was closed) and then at the University of Washington, where he taught from 1951 until his death. In 1974, President Nixon appointed him to the Board of Directors of Amtrak.

The Chicago School's view of the city has some important implications for how we view the urban as a scale. Perhaps the most obvious is that, at least in its earliest incarnations, the School's models treated the city in areal fashion. Moreover, the urban was seen to be very much an organic whole – the various bits of the city into which it was internally divided were understood to be connected together corporeally as part of an integrated urban scale, one whose spatial extent pulsated and rippled across the landscape as it spread out from the original CBD. Given human ecology's biological origins, the city was also seen in bodily terms, with its outer ring viewed as an epidermis enclosing the city's various interconnected neighborhoods – a fact which raises questions about intersections between views of the body as scale and the urban scale. Thus, in both the Park *et al.* model and the Hoyt version, the urban was primarily conceived of as an internally differentiated areal unit, one made up of discrete concentric circles or circles mutated into sectors as a result of the transportation network. Although Harris and Ullman abandoned the concentric circle model for a multi-nucleated model, they too viewed the city in areal terms as a collection of highly integrated absolute spaces, with the city itself understood as one large container of urban processes.

Certainly, the Chicago School's members recognized that a city's growth and structure are affected by the broader social and economic structure within which it is situated, but they nevertheless did largely see cities as fairly autonomous spatial entities, representing a distinct spatial arena that stood within a particular spatial hierarchy some- where between "the regional" (considered larger than/above it) and "the body" or "the neighborhood" which it contained. For such writers, Chicago was essentially a natural container wherein could be observed urban processes at work. More importantly, Chicago was taken to be the paradigmatic example of urban development, at least until the rise of the so-called "Los Angeles School" in the 1980s (*Urban Geography* 2008). Consequently, it was generally assumed that Chicago's specifics could be used to explain many other cities' urban development patterns. The result was that the Chicago School and those who adopted its tenets presented this, their urban exemplar, as an entity that could effectively be disentangled from the broader web of socio-economic and political relationships at work within the northern Great Plains such that, to all intents and purposes, it could be laid over any other part of the economic landscape as a template to explain urban forms. This conceptual construction reinforced views of the urban as a distinct and discrete unit or scale of social life.

Central Place Theory and the mathematical modeling tradition

If the ecological model and the Chicago School's ideas served as the basis for thinking about the urban in the early to mid-twentieth century, in the 1960s Central Place Theory (CPT) came to dominate geographers' thinking, although CPT and the urban ecology of the Chicago School were not seen as necessarily being at odds – CPT largely focused upon the relationships between cities whereas urban ecology sought to understand land-use patterns within them. Initially developed by German geographer Walter Christaller (1933/1966) to explain southern German settlement patterns, CPT argues that diverse types of urban areas develop in response to differential demand for various types of goods – whereas "low order" goods (milk, bread) are usually found in small settlements, "high order" goods/services (open-heart surgery, a university education) are typically only available in larger ones, because they require a bigger customer/user base. Consequently, there is an inverse relationship between settlement size and number. Likewise, the larger a settlement is, the farther away is one of similar size, whilst a settlement's range (how far people will travel to it) increases as its population increases. Based upon these assumptions, Christaller showed that different-sized settlements (hamlet, village, town, city, and regional capital) are organized in a unidirectional hierarchy, based upon whether they sell high- or low-order goods – a settlement providing high-order services will be surrounded by those supplying low-order ones, though this does not work in reverse. He also noted that settlements could be categorized according to their functionality – market centers, transportation centers, administrative centers (see Figure 3.1 and Box 3.2).

Christaller's model was later revised by German economist August Lösch (1938; 1954), who added a focus on manufacturing (Christaller's model had only looked at urban areas as retail/service centers). However, whilst Christaller started at the top of the urban hierarchy and moved down it (he first focused on the largest market area and then turned to commodities with ever smaller market areas), Lösch started from the bottom and moved upward (he considered first the commodity with the smallest market area and then introduced additional commodities with successively larger market areas) (von Böventer 1963). Furthermore, whereas Christaller's approach left an urban landscape in which there was a hierarchy of distinct tiers or steps according to settlements' size and function, Lösch's scheme presented a landscape with a more con-tinuous distribution of settlements of different size and function – he did not assume, for instance, that larger central places subsumed the functions of smaller ones nor that cities of the same size necessarily performed the same functions. Subsequently, Isard (1956) and Berry and Garrison (1958) developed CPT further. Still others sought to use analogies from physics to explain settlement patterns, in particular the "Gravity Model"

Box 3.2 WALTER CHRISTALLER'S CENTRAL PLACE THEORY

Walter Christaller (1893–1969) was a German geographer who studied cities not as individual entities but in terms of how they fitted together into broader urban systems. His approach is termed Central Place Theory (CPT) and was first outlined in 1933. Although favorably inclined toward socialism as a youth, in 1940 he joined the Nazi Party and during the Second World War worked in the government's Planning and Soil Office, where he was in charge of planning for occupied Poland (Rössler 1989). After the war he joined the Communist Party and in 1951 became a deputy in a local council (Hottes 1983).

At the heart of CPT was Christaller's belief that each city or town (i.e., central place) would dominate a hexagonal-shaped hinterland, given that hexagons could be fitted together across the landscape without any overlap (which would not be the case if hinterlands were circular). Given that each central place which served as a seller of higher-order goods would therefore be surrounded by six neighboring lower-order places, these central places were assumed to take one-third of the customer base from each of the six surrounding hexagons. Hence, when settlements served as *market centers*, they were given the designation of "3", representing a score of "1" for itself and one-third of each of the six surrounding hexagons, for a total of "k = 3" $(1 + [1/3 \times 6])$. On the other hand, when they served as *transportation centers*, they were given the designation "k = 4" because, in this case, it was more efficient for each central place to service half of each lower-order hexagon surrounding it $(1 + [1/2 \times 6])$. Finally, when central places were serving as *administrative centers*, it was more efficient for settlements to be nested in such a way that smaller settlements were entirely within the sphere of influence of larger ones, hence their designation of "k = 7" $(1 + 6)$. In outlining CPT, though, Christaller made several simplifying assumptions, including that all regions have a flat and undifferentiated surface in which transportation is equally easy in all directions and across which there is an evenly distributed population (one that has similar purchasing power) and resources, that there is perfect competition in the system, that people will always purchase goods from the closest place that offers that good, and that low-order goods will typically be offered for sale in more places to which individuals do not have to travel very far whereas high-order goods will be offered for sale in fewer, larger places to which most consumers will have to travel greater distances. He also assumed that each settlement had an economic hinterland around it (its "sphere of influence") and that each required a minimum level of population (its "threshold population") in order to survive.

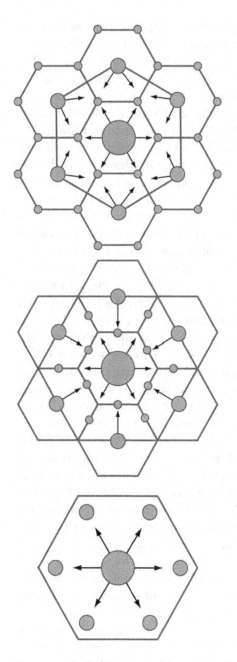

Figure 3.1 Idealized images of k = 3 (top), k = 4 (center), and k = 7 (bottom)
central places

which, paralleling Newton's gravitational equation, suggested that the degree of interaction between two settlements is related to their relative size and the distance between them (Zipf 1946; see also Reilly 1931).

Following this, Hall and Hite (1970) melded CPT with the gravity model to delineate economic areas as they sought to develop a three-fold hierarchy of central places in South Carolina based on commuter hinterland boundaries. Meanwhile, Smith (1976) suggested that Christaller's original model was only applicable to well-developed market societies and that in colonial societies, where exchange is tightly controlled, other patterns develop, which she described using different metaphors that have clear scalar implications. Hence, in the case of colonial Latin America, she detailed the existence of a networked "solar hierarchy" of hub and spoke relationships in which all low-order centers are focused on a single regional central place whose administrative power is such that they can shield themselves from competitors in nearby regions, whereas in cases where external powers control exchange a "dendritic hierarchy" may develop in which each lower-order central place is oriented to only a single higher-order center. Significantly, Smith (1976: 34, Figure 6) used a tree-like image to illustrate this latter situation. More recently, Meijers (2007b) has argued that rather than seeking to refine CPT, a more appropriate model of urban spatial organization, particularly in the context of polycentric urban regions, is that of the network. Drawing from Pred (1977), who pointed out that cities' functional hierarchies are often not consonant with their size hierarchies, such that relationships between cities may be not only vertical but also horizontal in nature, Meijers suggested that a network model better explains spatial reality and avoids problematic notions of geographical hierarchy – an approach which perhaps has shades of Marston et al.'s (2005) flat ontology. This is particularly the case in which there exists a group of similar-sized and politically discrete cities that function economically as a single urban unit but whose local specialization makes it difficult to argue that there is any degree of functional hierarchy between them.

In returning to issues of geographical scale, the most significant element in CPT is the fact that different types of settlement are generally defined by their functionality as they interact within a highly ordered and nested hierarchy – a focus upon order which, if we are to believe Sibley (1998), reveals almost a degree of psychosis in Christaller's need to find spatial regularity in the landscape. Thus, the urban landscape is divided into a series of absolute spaces and scales of uniform shape (hexagons) arranged according to size and function, with each level of the hierarchy providing different and distinctive services. Moreover, Christaller's model presents a landscape that is decidedly stepped in its scalar structure – a nested hierarchy of centers and markets evocative of the Matryoshka dolls metaphor – with different-sized settlements (k = 3, k = 4, etc.) serving as distinct spatial tiers within the overall landscape. Although Lösch's version of CPT is much less hierarchical and tiered (and in that sense the urban landscape is much less like a ladder, wherein one climbs up the various tiers/rungs from bottom to top, than it is, perhaps, like an escalator), in scalar terms it nevertheless shares many similarities with it – both have an areal view of scale, for instance. Significantly, in this regard both Christaller's and Lösch's models have similarities with Smith's (1984/1990) Marxist conception of the urban scale, with all three based upon the proposition that the urban is defined by the flows of people into and out of various central places (whether on a daily commuting basis or, perhaps, a weekly/monthly shopping basis). Nevertheless, in other regards they are quite different from Smith's understanding of how the urban landscape comes to look how it does. Hence, whereas for Christaller and Lösch the urban landscape is essentially static, for Smith it is highly dynamic, reflecting the deep structural forces of capitalist accumulation. Equally, neither Christaller nor Lösch saw scales as produced out of struggle. Instead, they saw the urban scale as algebraic and deterministic, with scales emerging from the logic of economic forces as interpreted and represented through mathematical formulae. Finally, whereas Smith argued that the urban scale emerges out of the dynamics of production, for Christaller it was shaped largely by the dynamics of consumption – if for Smith (1984/1990) Travel-to-Work Areas mark the urban scale's spatial extent, for Christaller it is different types of goods' ranges that do so.

Political economy approaches to the urban

In the 1970s, Geography's quantitative revolution started to wane and a new set of theories about the urban and urban form came to dominate. Although some drew on humanistic traditions to explore how people interpret the urban landscape (e.g., Ley and Samuels 1978; Duncan 1978), this humanistic tradition did not really theorize how the urban scale comes to be and with what import. These latter questions were important ones for Marxian political economy, which generally saw the urban as a central element in how capitalism functioned. For instance, Lefebvre (1970/2003; 1973/1976) argued that space's production in particular ways was central to capitalism's survival, that the urban played an important role in this, and that each mode of production generated its own type of urban space (Burgel et al. 1987). Indeed, capitalism, Lefebvre suggested, had undergone an "urban revolution" in which the source of accumulation had gradually shifted from its eighteenth-century industrial base to a twentieth-century urban one as the "the urban problematic [had] becom[e] predominant" (Lefebvre 1970/2003: 5). This shift had brought with it a transformation in how cities are organized spatially and also a transformation in concepts of the urban. Hence, using an interesting scalar image, Lefebvre (p. 167) suggested that "[u]rban society can only be defined as global[, for] it covers the planet." Significantly, Lefebvre did not equate "the urban" with the physical dimensions of the city, suggesting instead that "the urban revolution creates an urban society, in which case the physical separation of city and countryside becomes of less and less significance" (Saunders 1986: 158) (Box 3.3).

For his part, Castells criticized extant urban theory – specifically that of the Chicago School – for its failure to see urban forms under capitalism as having a particular historical specificity and set out to explore instead how urbanism took on particular forms and functions under capitalism which were different from the role played by cities under other modes of production (e.g., feudalism) (Brenner 2000). He also drew on Althusserian structuralism to critique Lefebvre for an alleged utopianism that, Castells suggested, did not recognize sufficiently the structural forces which shape urban forms and which saw the urban as a somewhat self-evident object of analysis, an ideological – rather than material – construction. In contrast, Castells distinguished two basic dimensions of the urban: its scalar and its functional aspects – whereas the former "concerned the materiality of social processes organized on the urban scale as opposed to supraurban scales," the latter "concerned not merely the geographical setting or territorial scope of social processes, but their functional role or 'social content'" (Brenner 2000: 363). Describing geographical scales as the "spatial units" into which the landscape is divided, Castells (1977: 444) saw the urban as the sphere of reproduction under capitalism and the regional as the sphere of production. Thus, for him (1978: 171) cities were "units of collective consumption" into which "the state increasingly intervenes" (1978: 3), such that the

Box 3.3 BRIEF BIOGRAPHIES OF HENRI LEFEBVRE, LOUIS ALTHUSSER, MANUEL CASTELLS, AND DAVID HARVEY

Henri Lefebvre (1901–91) was a French sociologist and Marxist who argued that what he called the "production of space" was integral to how capitalism as an economic system develops. After graduating in 1920 with a degree in Philosophy from the Sorbonne in Paris, Lefebvre worked with a number of other philosophers to try to transform the field. In 1928 he joined the French Communist Party (PCF) and was a professor of Philosophy from 1930 to 1940, when he joined the French Resistance. In 1958 he was expelled from the PCF and soon thereafter became a professor of Sociology at the University of Strasbourg. In 1965 he joined the faculty at the new university in suburban Paris at Nanterre, where many students would be involved in the social upheavals that shook France in 1968. He is perhaps most well known for his books *The Survival of Capitalism* and *The Production of Space*. Lefebvre was a critic of the kinds of structuralist Marxism advocated by Louis Althusser.

Louis Althusser (1918–90) was a French Marxist philosopher who was born in the French colony of Algeria but was educated in France. He was also a member of the PCF. He was ambivalent about the party's de-Stalinization initiated in 1956 and criticized the student movement of 1968 as an "infantile leftism," though he subsequently moderated his views. He suffered periods of mental illness during his life and in 1980 murdered his wife. Academically, he is perhaps most famous as a leading light of the French structuralist Marxism tradition, which argued that the principal function of the state and its institutions under capitalism is to reproduce the capitalist system. This occurs somewhat independently of the wishes of those individuals who are in government. The state as an institution, Althusser argued, therefore has a certain degree of autonomy from those who make up the capitalist class. This was a different perspective from a more "instrumentalist" approach, which argued that the state is directly responsive to the wishes and whims of the capitalist class. His most famous book is probably *Reading Capital*.

Manuel Castells was born in 1942 in Spain. As a young man he was active in anti-fascist politics, which caused him to have to flee

continued

Box 3.3 continued

Franco's Spain. He completed his studies in Paris and received a doctorate from the Sorbonne. In 1967 he became a professor of Sociology at the University of Nanterre. He subsequently moved to the United States, teaching at the University of California, Berkeley, and at the University of Southern California. His early works were on Marxist urban theory. In his 1972 *The Urban Question* he adopted an Althusserian stance and criticized Lefebvre's work for its alleged subjectivism. In the 1980s he began to explore the emergence of what he has called "the Network Society" and the "space of flows," which represents a new form of spatial arrangement under post-industrial networked capitalism.

David Harvey was born in Kent, England, in 1935. Educated at Cambridge University, he has spent most of his life working in the United States, principally at the Johns Hopkins University in Baltimore and, presently, at the City University of New York. Between 1987 and 1993 he was the Halford Mackinder Professor of Geography at Oxford University. His initial work was within the positivist tradition, as exemplified in his 1969 book *Explanation in Geography*. However, almost even before it had been published, Harvey moved away from positivism and adopted Marxism as a theoretical lens through which to understand the world. He outlined his initial ideas in this vein in his 1973 *Social Justice and the City*. This was followed by his 1982 *The Limits to Capital*, in which he sought to Marxify Geography and spatialize Marx. With his 1989(b) *The Condition of Postmodernity* Harvey defended Marxism from what he saw as a rejuvenated idealism masquerading as postmodernism. Harvey's work has been much influenced by that of Lefebvre.

urban could only be understood in terms of its function as a site for labor power's reproduction (1977: 236–37, 445).

In a rejoinder, Saunders (1986) argued that Castells was wrong to indicate that the kinds of processes that he suggested were occurring within cities (i.e., collective consumption and the reproduction of labor power) were specific to that geographical scale of social organization. Instead, Saunders saw cities' spatial structures as largely the contingent outcome of broader social forces and that "supraurban" processes are at work within them. According to Brenner (2000), at the heart of this debate between Castells and Saunders over whether or not the urban serves as a container for certain scale-specific processes (such as labor power reproduction) were two core assumptions. First, despite Castells's own criticism of Lefebvre for his failure to specify what the city is and how it serves as an object of scientific investigation (rather than existing as merely an ideological construct (Stalder 2006: 14)), both Castells and Saunders, Brenner averred, tended to view the urban as a rather self-evident entity – their questions concerned not so much how the urban scale came about but how it is represented. Second, Brenner contended, both Castells and Saunders tended to regard scales such as the urban as operating in a mutually exclusive – rather than co-constitutive – fashion with regard to other scales, in what Brenner called a "zero-sum" conception of scale. Hence, for Castells and Saunders, Brenner (2000: 364) suggested, supraurban scales (such as the national) "were merely external parameters for the urban question," rather than interdependent components thereof.

Other Marxist writers likewise explored the connections between the urban and capitalism in the 1970s and 1980s. Harvey (1973), for instance, engaged with Lefebvre's works but initially pondered whether the latter had gone too far in focusing upon conflicts over space, intimating that, perhaps, Lefebvre had fetishized space. Before long Harvey (1976, 1982, 1983) had accepted the notion of a "socio-spatial dialectic" and developed a detailed exposition of how capitalists create urban environments as central elements in the accumulation of capital and how such environments subsequently shape the possibilities for further capital accumulation, suggesting that "[c]apital accumulation and the production of urbanization go hand in hand" (Harvey 1989a: 22). Walker (1981), meanwhile, contended that the built environment's postwar transformation in the US, especially with regard to cities' dramatic physical spread, was part of a "suburban solution" wherein the federal government had encouraged building so as to stimulate the economy and profit rates. As mentioned in Chapter 1, for his part Smith (1984/1990) also argued that the urban is central to how capitalism operates and that the urban scale should be defined by the spatial extent of daily labor markets, as expressed through Travel-To-Work Areas. Although this approach presented a processual view of scale, it did, however, generate some issues that bear pondering. First, using TTWAs as the basis for determining the urban scale assumes that there is some physical movement of workers between two fixed points (their homes and locations of paid work). However, in the case of workers who telecommute, this is not the case and, as telecommuting likely becomes both more common and spans greater distances (sometimes stretching across international borders), using the TTWA as a definition of the urban scale may prove increasingly problematic, unless virtual travel is included.

Second, such a view could be taken to imply that the urban as a scale encloses the body as a scale, for people move around within a geographical area (the urban) whose boundaries are defined by the maximum commuting distance city residents can daily travel. Put another way, although its geographical area is defined by the circadian movements of commuters' bodies, the urban scale itself is seen to contain such bodies, for workers do not cross, on a daily commuting basis, the lines circumscribing their particular TTWA. This suggests a scalar hierarchy is at work (or at least envisioned to be so), although in an interesting development, Smith (1992b) showed how some of this spatial containment could be avoided by some bodies – paradoxically, those of homeless people, who may use "Homeless Vehicles" constructed out of modified shopping carts to move themselves and their belongings from place to place and thereby extend their bodies' spatial reach across the city. Third, because the type of paid work in which individuals engage and locations of residence vary greatly by gender, race, and class, the social average commuting distance for diverse types of commuters can differ significantly – in the US, women, racial minorities, and people in lower socio-economic groups typically have shorter commuting patterns than do men, whites, and wealthier people (Johnston-Anumonwo and Sultana 2006) – a phenomenon seen in other societies too (Blumen and Kellerman 1990). This means that an urban area's overall TTWA is usually defined by those who commute the furthest: typically wealthier, white males. There is, then, a built-in class, race, and gender bias in an otherwise apparently "objective" measure (the TTWA).

Meanwhile, feminists explored the urban as an entity through the lens of gender, focusing particularly on the long-standing division – both in urban theory and in urban policy – between public and private spaces. For instance, Hayden (1980) appraised how patriarchal relations had shaped urban development and pondered what a non-sexist city would be like, whilst Saegert (1980) critiqued views of the urban which linked the city with men and the suburbs with women. For her part, McDowell

(1983) suggested that the urban is a key spatial scale through which gender is experienced and constituted. Drawing upon such ideas, much of this early feminist work argued that because cities (at least those in the West) are usually built on the basis of some kind of demarcation between "productive" and "reproductive" spheres, the urban – or, more precisely, the suburban, given that women have generally been associated with the spaces of consumption, reproduction, and the private, all elements taken as emblematic of the suburban – represents the principal spatial sphere for women's everyday lives (Fincher 1990; McDowell 1993; Bondi and Rose 2003). However, although this work spoke to the spatialization of gender relations through the growth of suburbs in Western countries, because it associated women with reproduction (both biological and social) and the private space of the home (Marston 2000), it tended to focus upon the experiences of heterosexual, middle-class women in nuclear families. It largely ignored how different urban forms enable or constrain the expression of various sexualities, as well as the roles gay communities play in shaping the built environment – as through gay gentrification in particular urban neighborhoods (Castells 1983). More recently, however, this lacuna has begun to be filled, with several authors showing how the production of urban space – and thus the urban scale – is not neutral with regard to sexuality but is heavily shaped by it (Lauria and Knopp 1985; Knopp 1992; Forest 1995; Brown 2000). Finally, Herod (1994) suggested that many Marxist writers had tended to focus upon the internal workings of capital in seeking to understand the genesis and dynamics of the urban scale and that this left out how workers had historically shaped its form – the US labor movement, for instance, was heavily involved in encouraging post-1945 suburbanization.

Whereas some writers sought to explore the importance of the urban for how capitalism operates, others took a slightly different view, one which suggested that the urban is not, in fact, quite so central to the capitalist mode of production. Hence, Giddens (1981: 147) argued that with the advent of capitalism, the city was "no longer the dominant time-space container or 'crucible of power'" it had been under feudalism. Rather, this role has been "assumed by the territorially bounded nation-state," a development which, if we accept Giddens's argument, has important scalar implications. Thus, Giddens argued that the transition to capitalism heralded three fundamental shifts concerning the urban: 1) the city has increasingly been replaced "by the nation-state as the dominant power container" (pp. 147–48); 2) the contrast between the city and the country, the urban and the non-urban has been progressively eliminated within capitalist societies, such that there are fewer sharp divides between the absolute spaces of the city and of the rural; and 3) the social processes affecting cities in capitalist societies are, by and large, quite different from those affecting cities in non-capitalist societies. "Taken together," Giddens contended (p. 148), "these [shifts] represent a profound discontinuity between the city in class-divided societies and capitalist urbanism," a discontinuity whose most tangible physical expression "is the disappearance of the city walls, the physical enclosure of the [pre-capitalist] power container."

Generally, however, much critical analysis of the urban at this time increasingly focused upon what Brenner has called its "scale-specificity." Hence, whereas Castells's early conceptions of scales such as the urban saw them as spatial expressions of broader social functions, subsequently the "analytical core of the urban question [became] no longer the functional unity of the urban process but rather the role of the urban scale as a complex geographical materialization of capitalist social relations" (Brenner 2000: 364). In this regard, one important set of considerations involved how deindustrialization was impacting the urban and how urban areas were consequently conceived of by those seeking to understand the dynamics and material manifestations of this broader capitalist restructuring. This questioning of the place of the urban within such restructuring was known as "the localities debate" and was cast in the context of a wider questioning of Marxian theory's role in understanding the urban.

The localities debate

The localities debate arose in the mid-1980s in response to the Changing Urban and Regional Systems in the United Kingdom (CURS-UK) initiative, a program of research funded by the Economic and Social Research Council. The CURS initiative was itself designed to understand the effects of economic and social restructuring in seven "localities" in Britain: East Liverpool (a long-standing branch plant economy affected by chronic unemployment); Middlesborough (a center of the steel, chemicals, and health service sectors in the 1970s which experienced significant job losses in the early 1980s); South West Birmingham (which had had a growing economy in the 1970s but whose auto industry suffered declines in the 1980s); Lancaster (a former industrial city experiencing growth in some low-level service occupations, such as tourism); Cheltenham (a growing financial services and clerical community west of London); Swindon (a focus of various high-tech industrial expansion); and the Isle of Thanet (an area dominated by retirement services) (Smith 1987). A team of researchers collected data on various elements of social and economic life in each of these localities, including the labor market, gender, ethnicity, housing provision, the planning process, work culture, and others. These data were designed to provide a detailed profile of each locality and how restructuring was impacting it.

Although the initiative generated huge quantities of data, it was not without its critics. Smith (1987: 62), in particular, suggested that the research reified the urban scale and focused upon "the uniqueness of each locality." The project was, he averred, "primarily about the localities in and of themselves rather than an attempt to understand the dimensions of contemporary restructuring as revealed by the experience of these localities." Consequently, he argued, the initiative represented an "empirical turn," one in which individual places were being studied "for their own sake" rather than for the researchers' ability to use them to "draw out [broader] theoretical or historical conclusions." He also contended that the initial way in which the seven localities were chosen was problematic. Much of his critique revolved around the issue of geographical scale and, particularly, the ways in which, he maintained, scale had not been adequately theorized when operationalizing the research program. In Smith's assessment, the CURS initiators had adopted a very unsophisticated view of scale because they had used as one of their primary criteria for choosing localities simply their population size – all contained between 50,000 and 200,000 people. Such a basis for choice seemed inappropriate for two reasons, with both of Smith's complaints relating to matters of scale. First, he claimed (p. 63), by choosing such large population centers it was "difficult to see how any meaningful comments [could] be made about the regional scale." Second, although Smith did not himself use the term, he argued that this way of choosing localities seemed to generate geographical units of study that were what Sayer (1984) has called "chaotic conceptions" (Box 3.4) – the fact that the localities under consideration were a mixture of self-contained urban units (such as Lancaster and Middlesborough) and of slices of larger urban conurbations (such as East Liverpool and South West Birmingham) meant that quite different entities (whole communities versus various slices of others) were being compared. The result was a situation in which, despite the fact that there was "an inherent incomparability between an entire urban area, more or less coherent in functional terms, and a sector of a larger urban mass, plucked from its surroundings," they were being considered equivalent. "[L]ike the blind man with a python in one hand and an elephant's trunk in the other," Smith declared, "the [CURS] researchers [were] treating all seven localities as the same animal."

One of the key issues to emerge in this was what Smith called the "Gestalt of Scale" (see Chapter 1). In making a case (p. 64) that "[a]s the scale of economic activity is transformed so too is the geographical scale at which regions are constituted," Smith declared that there was a problem in extant theorizing about scale as expressed through the CURS initiative, for researchers did "not yet have either a generally agreed upon language or a theory of the development of geographical scale" with which to comprehend the transformations they were studying. Consequently, because the CURS project had seemed to choose localities on the basis of "the nature of the localities themselves . . . rather than a theoretical perspective on urban and regional change," it had failed to consider seriously how "the very constitution of the urban and regional scales [was] being utterly transformed by the restructuring process." Instead, the initiative presented scale as little more than an optical effect – by choosing different-sized cities, the CURS researchers hoped to shed light on different urban processes but were ignoring how scales themselves were being materially reworked.

Not surprisingly, researchers involved in the initiative defended it from Smith's criticisms. Cooke (1987: 72) argued that "[b]eing empirical is by no means the same thing as being empiricist" and that "theory has a high status-value in CURS, it informs the research strategy and design of the programme, but it is being deployed diagnostically not dogmatically." Furthermore, he (p.77) recognized that "[b]ecause CURS focuses on localities which are, inevitably, singular and therefore unique, the fear is raised that the research will be idiographic, unable or unwilling to make what is perceived as the essence of intellectual enterprise namely, generalisations." However, he then went on to declare that the goal of the CURS initiative was not to be able to generalize across these various localities but, rather, to use "clinical inference . . . to seek generalisation . . . within cases." Hence, for Cooke, this approach was "peculiarly relevant to locality studies," for it was "the means by which a theoretical framework or construct about general interactions in localities might be

Box 3.4 CHAOTIC CONCEPTIONS

In his 1984 book *Method in Social Science: A Realist Approach*, Andrew Sayer outlined the importance of abstraction, which is the manner in which the objects of research are separated from the contexts within which they are found so that they can be examined. Abstraction is necessary because the world's complexity makes it impossible to study everything at once. We must study it bit by bit. However, in engaging in abstraction, it is important that we do not separate objects which can only exist together nor put together objects which can exist separately. For instance, a slave can only exist if there is a slave owner – the two are necessarily connected. On the other hand, there is no necessary connection between skin color and the status of being a slave. Thus, both blacks and whites can be either slaves or slave-owners. What results in one group actually being slaves and another not is nothing that is inherent to either but is, rather, the result of various contextual factors, such as how the legal system (which can designate one's status in this regard) operates and who controls that system. For Sayer (pp. 126–27), the key to understanding causal relationships is distinguishing between a *rational abstraction*, "which isolates a significant element of the world which has some unity and autonomous force" and a *chaotic conception*, which "arbitrarily divide[s] the indivisible and/or lumps together the unrelated and the inessential, thereby 'carving up' the object of study with little or no regard for its structure and form." Sayer argues that the concept of "service employment" is a chaotic conception because it puts together various activities which "neither form structures nor interact causally to any significant degree," such that many service sector jobs "lack anything significant in common."

developed." The approach adopted by the CURS initiative would be to "generalise within cases and compare those generalisations, to the extent it becomes possible, with generalisations drawn from within other cases." This approach, Cooke maintained (p. 78), addressed some of the scalar matters that Smith had raised because it was "conscious of the need for vertical and horizontal conceptualisation" and asked "questions about both locality and process." Thus, he contended, whereas "[s]ome processes are primarily local – neighbouring, community involvement, aspects of cultural practices, political organization, services consumption and occupation-specific employment . . . [o]thers are less so, notably those having to do with the ownership and production of most traded services and manufactures, cultural provision through the mass media, and governmental sovereignty." In the end, Cooke maintained that Smith's notion of "the Gestalt of Scale" merely seemed to "obfuscate the real theoretical and empirical question of precisely what aspects of a process such as suburbanization . . . derive from local factors such as spatial variation in demand, and which from national and international factors such as the price of credit."

Several issues emerge from this discussion of the urban scale's status. One concerns the nature of observation and the relationship between how the world is imagined to be and how it actually is. In particular, Smith's position was that, regardless of how changing one's point of view may change one's understanding of material processes, this does not change the ontological nature of scales which, for Smith, exist as real, coherent things. Indeed, this is why he had such problems with the CURS researchers' practice of focusing upon particular slices of larger conurbations and then treating them as if they were largely self-contained entities and equivalent to localities which are fairly whole urban units and which could be seen to have a certain degree of scalar coherence, based upon the spatial extent of their local labor markets. Consequently, whereas studies of Lancaster and Cheltenham focused upon entities which were largely defined by their TTWAs' geographical extent, East Liverpool and South West Birmingham represented swathes of built environment effectively ontologically ripped from the broader urban conurbations within which they sit and across whose boundaries myriad people travel on a daily basis – say, from northwest Birmingham to southwest Birmingham – to get from home to work and vice versa. For Smith (2000), the CURS initiative had failed to distinguish between "methodological scale" (created by the researcher through his/her choice of focus) and "geographical scale" (created by material processes at work in the landscape).

Perhaps the most significant issues with regard to scale, though, relate to what the way in which localities were conceptualized by both sides in the debate says about their understandings of scale. In particular, despite their differences both Smith and Cooke viewed localities as areal units – as "platforms of absolute space in a wider sea of relational space"

(Smith 2000: 725). Hence, whereas Smith (1987: 63) questioned the CURS initiative for appearing to cut off analytically individual localities from broader processes, seeing localities as either "self-contained urban units" or as "slices of much larger conurbations" – both areal scalar representations – Cooke (1987: 77) used areal language when defending the project by arguing that clinical inference was about seeking generalization "within" rather than "across" case studies (that is to say, "localities"). Likewise, Cooke's (p. 78) call for "the need for vertical and horizontal conceptualisation" seemed to reify both verticality and horizontality in a manner decried by Marston et al. (2005, 2007; also, Jones et al. 2007). The result is that much of the "localities debate" (Cooke 1989; Harloe et al. 1990; Cox and Mair 1991; Swyngedouw 1997b) saw scale in areal terms, with scales providing the boundary around various segments of absolute space but with each such areal unit understood to be located within a vertical hierarchy of areal scales that stretch upwards to the national and global scales.

THE URBAN AS NETWORK

Whereas much twentieth-century theorizing has viewed cities in areal terms, in recent years a non-areal way of thinking about the urban, one influenced by the broader fascination with network theory, has emerged. Although Vance's (1970) theorization of urban development has sometimes been seen as a "network model" (e.g., Conzen 1975; Pacione 2005: 134), and Bohannan and Bohannan (1968: 194–219) had characterized the central market places of Nigeria's Tiv ethnic group in terms of networks (Smith 1976: 39–44), it was only really in the 1990s that the language of networks came to play much of a role in considerations of the urban. Amongst the first to argue in any systematic and critical fashion for a network model to understand the urban were Hohenberg and Lees (1985), who suggested that the historical development of European urbanism might in some ways be better represented by a network analogy than through CPT. Hence, they maintained that although CPT has some historical explanatory value, it fails to explain adequately the distribution of urban areas in the medieval period. Thus, whereas CPT "implies a more or less even spatial distribution of cities around a central capital, with regional boundaries typically falling in zones of weak urban interaction," in actual fact in regions such as Picardy or the Po plain, regional boundaries ran through the middle of quite active zones, whilst cities such as Cologne, Toledo, and Prague acted "as gateways or portals into their region, not as distance-minimizing central places" (Hohenberg and Lees 1985: 58). Moreover, whereas CPT views urbanization as the result of rural development, with cities serving as the focal points of regional economic growth, many large European cities developed not because they were surrounded by productive agricultural lands but because they served as the points of export of staples and imports of luxuries. Consequently, Hohenberg and Lees averred, focusing not on agricultural hearths but upon trading links provides the basis for a better model of some aspects of European urbanization than does reliance upon CPT. In this model – which they called the "Network System" – cities should be seen to "form the centers, nodes, junctions, outposts, and relays of the network" (p. 62) (see Table 3.1).

Others also began to examine critically the idea of networks. For instance, Camagni and Salone (1993) argued that although in traditional regional science understandings of urban landscapes the landscape is sometimes viewed as consisting of a network of cities, this network is typically structured in a highly hierarchical way – as in the case of Christaller's CPT. However, they suggested, whereas Christaller's theory may have been appropriate for the time in which he developed it, by the end of the twentieth century city systems in increasingly post-industrial societies had come to depart ever more from the Christallerian pattern of a nested hierarchy of centers and markets, largely because the "reduction in transport costs and the demand for 'variety' by consumers have broken the theoretical hypothesis of separated, gravity-type, non-overlapping market areas" (p. 1055). Instead, "synergystic [sic] elements operating through horizontal and vertical linkages among firms have generated the emergence of specialised centres, in contrast with the typical despecialisation pattern deriving from the theoretical model[, and] high-order functions [often now] locate . . . in small (but specialised) centres where the model's expectations refer only to lower-order functions." Arguing that "in the Christallerian and Löschian models, the shape of the urban hierarchy was determined by the interplay of forces like economies of scale, minimum efficient production size, demand density and market size," Camagni and Salone suggested that "other production forces working at the micro-economic and micro-territorial scale may be considered as the driving forces of the new 'network' paradigm."

Taking this further, Camagni and Salone (p. 1057) contended that city networks are "systems of horizontal, non-hierarchical relationships

Table 3.1 Characteristics of central place and network systems

Characteristic	Central place system	Network system
Structure		
Basic unit	Agricultural region and/or local administrative unit	Trading network
Role of city	Central place in a hierarchy of central places	Node in a network of linked cities divided into core and periphery
Shape	Territorial and geometric	Maritime and irregular
Ideal type	Market city (Leicester)	Merchant city (Venice)
Functions		
Economic	Marketing and services	Trade, particularly long distance
Political	Administrative with hierarchical, regular links to local units	Informal controls or imperial hegemony
Cultural	Orthogenetic	Heterogenetic
Evolution		
Economic pressures for growth	Supply-push	Demand-pull
Direction of development	Up from the base	Out from the core
Prime movers	Producers: craftsmen and farmers	Traders: merchants and bankers

Source: Hohenberg and Lees (1985: 65)

among specialised centres, providing externalities from complementarity/ vertical integration or from synergy/co-operation among centres," with "the foundations for th[is] . . . concept . . . derived from the logic of the firm in its spatial behaviour," especially its networked linkages to other firms located at some distance from it. On this basis, they determined there to be three types of city-networks. First, there are "hierarchical networks," in which particular cities control certain market or production areas – as with the sub-contracting areas of large corporations around big cities such as Toyota City or Turin. Second, there are "complementarity networks," which are based upon economies wherein there is a great deal of vertical integration, with such networks made up of specialized and complementary centers that are interconnected through market interdependencies and the inter-urban division of labor is sufficiently broad that it assures a market area wide enough for each center – as with the Padua–Treviso–Venice area in Italy, where there is a fairly congruous division of urban functions between the main centers. Third, there are "synergy networks," which are based upon network externalities and are constituted by similar and co-operating centers in which the network integrates the market area of each center. Examples of these include financial cities, whose markets are tightly integrated via high-speed telecommunication media, and tourist centers, where all of the urban areas within a particular area are often included in cultural or historical itineraries – as in the cities and towns of, respectively, the Nord-Pas de Calais (France) and Wallonia (Belgium), where a cross-border initiative to develop common infrastructure has been encouraged by the European Union, and the Loire Valley.

Other scholars have likewise argued for a "network model" of the urban. Drawing upon Hohenberg and Lees (1985), Batten (1995: 320) explored the development of network cities in the Netherlands and Japan, contending that "it is network cities that have accounted for an above-average share of urban growth in today's Europe . . . [and that a]lthough some larger cities possess both network and central place characteristics, it is the smaller network cities that have counteracted the central place trend towards [urban] primacy and contributed to the size-neutrality of urban growth." Massey (1993: 148), in a new take on the localities debate, proposed that localities should be conceived of not as areal units with tightly drawn boundaries around them but, instead, as "nets of social relations." Camagni (1993: 66), meanwhile, maintained that by the 1980s the "real city-systems in advanced countries ha[d] deeply departed from the abstract Christallerian pattern of a nested hierarchy of centres and markets," the result largely of new, decentralized, and networked "post-Fordist" (Amin 1994) ways of organizing firms (in contrast to the highly vertically integrated and geographically concentrated form of industrial organization of the Fordist era). Consequently, he recommended, urban geographers should draw upon the concept of "firm networks" and develop an analogous concept – that of "city networks" – to understand how contemporary developments are shaping the evolution of urban systems. In a significant shift from his earlier position, Castells (1996) also came to define cities as networked phenomena, maintaining that cities are nodes in the "space of flows" and that the most important element within the urban system is the structure of the network, rather than the roles that particular cities play within it. For his part, Murray (2000: 3) adapted the model of "Artificial Neural Networks" as a heuristic device to explain the nature of trade in the fourteenth-century European urban network and, specifically, the role played by Bruges (Belgium) in it, arguing that the complexity of Bruges's merchants' behavior "simply cannot be approximated by static or purely hierarchical models (Bruges as *gateway*)." Whereas, however, Murray merely used the analogy of the human brain and nervous systems to explore Bruges's role in Europe's urban network, in his ponderings about the urban Virilio (1993, 1997), in an interesting merging of the bodily and the urban scales, intimated that under contemporary capitalism the urbanization of time is consonant with the urbanization of space, in that urbanization dramatically speeds up the tenor of modern life. For Virilio (1997: 20), then,

> urbanization of real space is . . . being overtaken by th[e] urbanization of real time which is, at the end of the day, the urbanization of the actual body of the city dweller, this *citizen-terminal* soon to be decked out to the eyeballs with interactive prostheses based on the pathological model of the 'spastic', wired to control his/her domestic environment without having physically to stir.

The body, in other words, is networked into the broader urban fabric, such that it is impossible to say where one ends and the other begins.

Rather than replacing CPT with a network model, these developments generally simply merged them together. Hence, Hohenberg and Lees (1985: 6) suggested that both the CPT and the Network System model have their place when explaining historical patterns of European urban development, as the two signify "contrasting modes of development, from the rural base upwards in one case and from the urban core outward in the other" – a ladder and a concentric circle image, respectively – and the experience of most European countries "shows both sorts of processes at work, implying that many cities, notably large ones, have a place in both sorts of systems." In examining more contemporary situations, on the other hand, Camagni and Salone (1993) and Batten (1995) assumed that CPT was most typical for industrial economies whereas the network model was more applicable to service sector-dominated economies. In contrast, however, Meijers (2007a: 9) indicated that the two ways of thinking about urban systems were so incompatible that seeking to amalgamate them made little sense:

> [whereas] the central place system emphasises, amongst other things, centrality, size dependency, a tendency towards primacy, a dominance of one-way flows, a fixed number of spatial scales, economic functions raising with and linked to scale and an even territorial distribution of urban population, the network model, on the contrary, emphasises nodality, size neutrality, a tendency towards complementarity, two-way flows, a variable number of spatial scales, variable sets of functions on the same scale and an uneven territorial distribution of urban population.

Accordingly, based upon the empirical evidence he had surveyed with regard to the hospital sector in the Netherlands, he pondered whether, rather than a merging of these two models, there was instead a new paradigm of spatial organization based upon networks that was replacing the central place model, at least in polycentric urban regions (Meijers 2007b).

This changing paradigm, from an areal and ladder-like hierarchical view of the urban to a flatter, networked one, had important consequences for how the urban as a scale came to be thought of by many in the 1990s. As Brenner (2000: 366) has put it: "[i]n contrast to previous conceptions of the urban as a relatively self-evident scalar entity . . . urban researchers [were increasingly] confronted with major transformations in the institutional and geographical organization not only of the urban scale, but also of the supraurban scalar hierarchies and interscalar networks in which cities are embedded." As a result, urban researchers increasingly began to conceptualize the urban scale with direct reference to various supraurban re-scaling processes. For Brenner, the outcome was a situation in which whereas "the urban question had previously assumed the form of debates on the functional specificity or scale-specificity of the urban within relatively stable supraurban territorial configurations, [by] the 1990s the urban question [wa]s increasingly being posed in the form of a *scale question.*" Three issues dominated this "scale question." First, both urban and economic geographers increasingly called attention to the importance of place-specific social relations – as manifested, for instance, in arguments about glocalization's significance for corporate decision-making – as being at the heart of "global" economic transactions, wherein the urban serves as a local-ized node within global flows and the global is constituted through networks of localities and cities. Second, analysts saw dramatic shifts in both the vertical and the horizontal relations among cities – as with the emergence/consolidation of new global urban hierarchies (vertical) and accelerated informational and financial flows between cities (horizontal). In such views, the urban was seen not only as a nested level within broader territorial hierarchies but also as the outcome of dense interscalar networks which linked widely disparate geographical locations. Third, Brenner (p. 366) indicated that the regulationist-inspired analyses of the 1980s and 1990s had linked contemporary urban restructuring to various transformations of the nation-state which were shifting power to both supranational and subnational forms of governance, such that the urban scale had increasingly become "not only a localized arena for global capital accumulation, but a strategic regulatory coordinate in which a multiscalar reterritorialization of state institutions is currently unfolding." These considerations of the urban question as a scale question were most particularly manifested in the growing "global cities" literature, a topic to which I return in Chapter 6.

CONCLUDING REMARKS

Several issues emerge from the above discussion. Perhaps the most important is that the forces which produce the urban scale have varied historically, depending upon the extant mode of production. Hence, in feudal times the city was principally a place for market exchange whilst under industrial capitalism it increasingly became a locus of waged labor, with its spatial extent delineated by the daily migration of labor within

TTWAs rather than the migration of commodities from the countryside into its walls. Equally, whereas in feudal Europe the towns were the material expression of a growing class divide, in ancient times cities such as Rome never played an important economic role with regard to their hinterland and the city served a merely administrative function in terms of its relationship with the surrounding countryside – as Marx and Engels (1845/1970: 90) wrote, "Rome . . . never became more than a city; its connection with the provinces was almost exclusively political and could, therefore, easily be broken again by political events," a situation in stark contrast to the emerging functional division between town and countryside represented by the medieval town. Weber and Durkheim likewise indicated that the urban's importance has changed over time and across space. Hence, as Saunders (1986: 38, 49–50) put it:

> Just as Marx denies the theoretical significance of contemporary urbanism on the grounds that the town-country division no longer expresses an underlying class contradiction, so Weber denies it on the grounds that the city is no longer a meaningful and autonomous unit of economic and political association . . . [Equally,] Durkheim does not consider the modern city relevant to the key concerns of social theory in advanced capitalist societies. Like [Marx and Weber], he argues that it is only in the Middle Ages that the city was significant in itself since it was only during that period that it provided the organizational expression for functional economic interests. Just as Marx and Weber see the city in antiquity as theoretically unimportant, so too does Durkheim, arguing that . . . the basis of association [in cities such as Rome] was familial rather than urban. And just as Marx and Weber deny the theoretical significance of the modern city (since for Marx it no longer expresses essential class relations, and for Weber it is no longer the basis for human association), so too Durkheim argues that the distinction between the city and the society as a whole in the modern period is no longer meaningful, that the society itself can now be likened to one great city.

Giddens, likewise, averred that the urban's role has not been historically or geographically constant. This recognition has critical consequences, for it means, amongst other things, that the urban as a scale of social organization has been more or less important, relative to other scales, at different times in different places.

A second set of considerations to emerge from the discussion above concerns how understandings of the urban vary historically and geographically. Thus, as we have seen, the urban meant particular things to the writers of the Old and New Testaments, things which were different from what it meant to the early Greeks or Romans. Likewise, it meant different things to the Maya and the Conquistadors, and different things under feudalism and under capitalism, with such different meanings often reflected in literatures of various periods and places (Mumford 1961; Pike 1981). Meanwhile, more contemporaneously, Castells (1977) argued that the urban is the sphere of reproduction under capitalism whereas the regional is the sphere of production, a theoretical binary which has shades of the kind of functionalism put forward by Taylor (1981). The significant issue for our purposes here is to ponder how these different comprehensions of the city have shaped how the urban scale is theorized, especially in terms of its relationship with, or importance relative to, other scales such as the national or the global.

Finally, there is the matter of how the urban as a scale is conceptualized spatially – variously as an areal unit/container within a broader hierarchy of scales or as a node within some kind of scalar network, for instance. As discussed in Chapter 1, the image of the network as a descriptor of scalar relations has certainly become more common in recent years and theorizations of the urban do not seem to be an exception to this trend. Hence, whereas the Chicago School and CPT were inclined to view cities as areal units, more recent views have often made use of a network metaphor, to the extent that Meijers (2007b), at least, has seen in this the emergence of a new paradigm of urban theory. Whether this change in metaphor relates primarily to material changes in how urban landscapes are constituted or whether it merely reflects a broader fascination with networks in the computer age is a more difficult question to answer, but the fact that the language being used to think about the urban scale appears to be changing from a topographical to a topological one is noteworthy. Relatedly, whereas theorists such as Weber tended to view the city as a somewhat separate arena of social life ("a partially autonomous organization" (1925/1978: 1220)), others (e.g., Smith 1995) have seen it as much more integrated into the broader gestalt of scale.

4

THE REGIONAL

While geography has much to offer in the way of systematic or topical analysis, . . . it is patent that the regional concept is its most valuable contribution.

Renner (1935: 137)

[W]hereas many regional systems have been devised by geographers there have been relatively few attempts to suggest any principles of regional division.

Grigg (1965: 465)

For much of the twentieth century in Anglo-American Geography – and other social and natural sciences, for that matter – the region was a key analytical concept. Indeed, for some it was Geography's central concept, the one which allowed the discipline to claim its privileged place as an integrative science or, even, as the "mother of sciences" (Hartshorne 1939: 373). However, arguments as to what, exactly, a region is and how regions should be delineated have coursed through the discipline during the past century or so. Thus, are regions "real" material things or instead structures imposed by the mind on the landscape? Are they given by Nature or constructed by humans? Are they coherent geographical scales or do they simply serve as some kind of spatial average of the extremes of "the generality of aggregate social structure . . . and the uniqueness of *locale*" (Cooke 1985: 213)? Or are they merely categorical devices which serve as a spatial equivalent to "the period" in history (Baker 2003: 159; Wishart 2004)? These questions have been particularly nettlesome, for of all the scales with which geographers have concerned themselves, it is probably fair to say that the region has been the scale which has been most frequently conceptualized in spatially rather vague terms – although they have generally been viewed as geographical entities that sit somewhere between the urban and the national scales in a hierarchy of spatial resolutions, regions have been seen to vary considerably in geographical extent, from rooms in houses (Giddens 1984: 119) to continental-sized areas.

At the same time, regions have been classified in several quite different ways. Hence, they have been defined in *functional* terms, an approach which assumes no degree of spatial homogeneity (as when a newspaper's sales area includes quite different types of place – rural and urban, wealthy and poor); in *formal* terms, an approach which does presuppose a degree of homogeneity across particular absolute spaces (as when describing areas of common cultural heritage); in *vernacular* terms, as when regions (such as "Dixie" in the US context) are thought to exist in the mind's eye; and in *nodal* terms, wherein regional cores have dominance over particular hinterlands (Whittlesey 1954; Nystuen and Dacey 1961). Equally, whereas formal and vernacular regions have typically been viewed in areal terms, with the region's border conceived of as a spatial circumscriber of more or less homogeneous spaces, functional and nodal regions have often been seen in networked terms as the product of interlinked social or natural relationships that criss-cross the landscape, though the two are not reducible to each other – as Symanski and Newman (1973: 350) note, "though nodal regions can be considered functional, functional regions, without further specification, cannot be considered nodal." Thus, whilst functional regions may be represented pictorially in areal terms for purposes of exposition, they are primarily understood in terms of the relationship between a core point and outlying nodes, all of which are linked through linear connections – for instance, the rail lines that connect a central city with the outlying suburbs to form a functional region can be thought of as forming a kind of skeletal frame over which the space of the region is draped.

In this context, this chapter traces the history of developments in the concept of the region as it has been employed within Geography. Although the chapter initially addresses developments in late nineteenth-century French and German Regional Geography, its primary focus is how understandings of the region have advanced in Anglo-American Geography. The chapter itself is broken into four main sections. The first outlines the growth of interest in regions in early twentieth-century Geography, especially as it emerged from the shadow of environmental determinism. In particular, it traces efforts to develop so-called "natural regions" and how humans were subsequently incorporated into considerations of regions. The second section explores how the regional scale was conceived of during the mid-century period when attention shifted to regions' human aspects, whereas the third recounts how the region was theorized during the so-called Quantitative Revolution of the 1950s–1970s. In the fourth section I cover efforts in the 1980s to develop a "Reconstructed Regional Geography" wherein a Marxist view of regions as material entities – in stark contrast to the Kantian idealist notion of regions as mental constructions – came to prominence. Finally, I make some broader observations concerning how narrative is structured and how this shapes our conceptualization of the world, and what this means for thinking about both regions and the way in which the intellectual history of Geography is often presented.

EARLY REGIONALIZATIONS

From Geography's earliest beginnings as a modern discipline in the eighteenth century, the region has served as an important concept. Although in classical times Eratosthenes divided the Earth into climatic zones and Ptolemy was interested in the Roman Empire's different cultural regions, one of the earliest modern writers to opine about regionalization was the German Johann Christoph Gatterer. A contemporary of Kant, Gatterer developed a system of regionalization that divided up the world into natural zones, with his work having a significant impact upon later geographers, including Alexander von Humboldt. Though interest in regionalization waned during the nineteenth century, especially as a "man-land" tradition (infused by environmental determinism (Livingstone 1992; Peet 1985)) came to dominate academic thinking, by the early twentieth century attention to regionalization had begun to grow again. By the First World War, regions were at the center of theoretical debate, despite conflicts over nomenclature – whereas Ernest Ravenstein advocated calling the study of regions "Länderkunde" or "chorography," Halford Mackinder, believing the term chorography not sufficiently "English" to "take root," favored Regional Geography (see Mill et al.

1905). Initially, these developments had their origins largely in France, Germany, and Britain, though by mid-century it would, arguably, be in the US where Regional Geography would find its most fertile ground.

In France, if a single individual can be identified with the idea of the region during the *fin de siècle*, it was undoubtedly Paul Vidal de la Blache, often considered the "father of French geography" (Buttimer 1971; Sanguin 1993; Clout 2003a) (Box 4.1). Influenced by sociologist Emile Durkheim's functionalism, Vidal de la Blache (1918/1926) proffered the idea of the "terrestrial unity" (*l'unité terrestre*) of the "Earth organism," in which the Earth was seen as an integrated whole whose various parts are fundamentally interrelated and obey general laws. This idea was perhaps most clearly expressed in his belief that geographers should not conceptually divide landscapes that Nature had materially put together. Adopting a position of environmental possibilism (Berdoulay 1981) and believing it important to develop a scientifically rigorous approach to studying regions, Vidal de la Blache (1903/1994) focused mostly upon how the identities of the various regions (*pays*) of France were shaped by the local cultures which had molded their natural environments over time. Although frequently seen to have been espousing a neo-Kantian view in which a region is simply a convenient classificatory device and not real in any ontological sense, more recent evaluation has suggested he may have had a more materialist understanding of regions (Archer 1993) – he identified, for instance, a hierarchy of "natural regions" (i.e., regions given by Nature) based upon France's physical geography within which a local *genre de vie* (mode of life) develops. Finally, espousing a Lamarckian organicist perspective toward geographic phenomena in which he attempted to "determine the processes by which humans and their natural environments . . . become . . . a 'complicated amalgam' in specific places" (Archer 1993: 499), he argued (e.g., 1913: 299) that there was such a terrestrial unity between humans and Nature that thinking about them in dualistic terms made little sense. This perspective was quite different from the environmental determinism then dominant in US Geography, wherein Nature was seen as separate from the humans whose behavior it determined (e.g., Semple 1911; though see Braden 1992: 242, who maintains that Semple was more structurationist (Giddens 1984) than determinist in her approach). Other French geographers (e.g., Fèvre and Hauser 1909) adopted similar Vidalian approaches to understanding the landscape.

Box 4.1 EARLY THEORIZERS OF THE REGION

Paul Vidal de la Blache (1845–1918) was born on France's Mediter-
ranean coast but educated in Paris. In the late 1860s he spent three
years in Greece studying archaelogy before returning to Paris to
complete a doctorate in 1872. He taught at several universities,
including the University of Nancy, the École Normale Supérieure, and
the Sorbonne, the latter both in Paris. He also helped found the *Annales
de Géographie*. In contrast to environmental determinists, Vidal de la
Blache argued for an environmental possibilism, which argued that
Nature places limits on human behavior but does not determine it
– as he put it: "There can be no question of a geographical determinism. Geography is nonetheless the key that one cannot do without"
(1909). He is considered by many to be the founder of modern French
Geography. Although influenced by Emile Durkheim, in many ways
the two were quite different – whereas Durkheim sought to establish
sociological laws, Vidal de la Blache focused on gaining a deep
understanding of individual regions, though he considered this a
potential basis for making more general inferences (Nir 1987).

Andrew Herbertson (1865–1915) was raised in Galashiels, Scotland.
He entered the University of Edinburgh in 1886, studying Natural
Philosophy, Physics, Geology, and Agriculture. He subsequently
studied at Montpellier in France, where he was influenced by Vidalian
ideas. In 1898, Herbertson was awarded a Ph.D. from the University
of Freiburg-im-Breisgau for his work on global rainfall patterns.
He taught at the Universities of Dundee, Manchester, and Oxford,
amongst others. At Oxford he was heavily involved in establishing
summer schools designed to better the teaching of Geography,
particularly at the pre-university level. Herbertson was a close friend
of the sociologist Patrick Geddes (himself influenced by French ideas
about regionalism), who introduced the concept of "region" into urban
planning and is credited with coining the term "conurbation" to
describe a regional urban agglomeration (Crone 1951: 47).

John Unstead (1876–1965) was born in London and became a pupil
teacher at the age of sixteen. He completed his undergraduate degree
in Politics and Economics at the University of Cambridge in 1898 and
then taught in an elementary school. In 1900 he saw Halford
Mackinder give a lecture at the London School of Economics and

continued

Box 4.1 continued

became one of Mackinder's students. His early work was on Canada's raw materials and their use for the British Empire, for which he was awarded a D.Sc. from the University of London. He taught at Goldsmiths' College and Birkbeck College in London, though took part in several of Herbertson's Oxford summer schools. Unstead wrote a number of school Geography textbooks. He was a founding member of the Institute of British Geographers and Chairman of the Geographical Committee of the League of Nations after the First World War.

Herbert Fleure (1877–1969) was born on the island of Guernsey, in the Channel Islands. He was educated at the University of Wales in Aberystwyth and spent much of his professional life there, as a professor of Zoology, of Anthropology, and of Geography. Fleure was later a professor of Geography at the University of Manchester. He was also a Fellow of the Royal Society and President of the Royal Anthropological Institute.

Alfred Hettner (1859–1941) was born in Dresden and completed his doctorate at the University of Strasbourg (then in Germany). Influenced by Kantian philosophy, Hettner conducted fieldwork in Chile and Colombia in the 1880s, and later in Russia, North Africa, and Asia. He spent most of his academic career at the University of Heidelberg and popularized the concept of Chorology (the study of the areal differentiation of phenomena and the relations between them within a particular region), largely through his teaching and publishing in the influential German geographical journal *Geographische Zeitung*. Hettner's notion of chorology is often seen to have had a significant influence on American geographer Richard Hartshorne's 1939 *The Nature of Geography*, though Harvey and Wardenga (1998, 2006) have suggested that Hartshorne significantly misread Hettner on several matters, such that many German concepts introduced to Anglo-American Geography *via* his interpretation of Hettner were done so erroneously.

Friedrich Ratzel (1844–1904), the son of the Grand Duke of Baden's housekeeper, studied Zoology at university. After completing his Ph.D. in 1868, he wrote a commentary on Charles Darwin's ideas, *Being and Becoming of the Organic World*. He joined the army during

continued

Box 4.1 continued

the 1870–71 Franco-Prussian War, after which he traveled in the United States, Cuba, and Mexico. He held academic positions in Munich and Leipzig, where he taught American geographer Ellen Churchill Semple, who was responsible for introducing into US Geography many of his ideas concerning Anthropogeography (the study of how the environment shapes human activities). Ratzel is perhaps best known, however, for popularizing the term *Lebensraum* ("living space"), which he borrowed from German biologist Oscar Penschel (who had used it in an 1860 review of Darwin's *Origin of Species* (Heffernan 2000: 45)) and which would be developed later by geographer Karl Haushofer and used by the Nazis to justify German expansion into Eastern Europe (Rudolf Hess, Adolf Hitler's deputy, was a student of Haushofer).

Albrecht Penck (1858–1945) was born near Leipzig and had interests in both Geography and Geology. After teaching at Munich University he was Professor of Geography at the University of Vienna (1885–1906) and at the University of Berlin (1906–27). At the latter he was Director of the Institute and Museum for Oceanography, as a result of which an oceanographic research ship – the "Prof. Albrecht Penck" – is presently named after him. Interested in both Geomorphology and Climatology, Penck conducted important work in the Bavarian Alps on the historical geography of its glaciations. He is noted for his work on the classification of climate and land forms and for refining Ratzel's concept of *Lebensraum*. He was elected to the Royal Swedish Academy of Sciences in 1905. Penck was the father of geomorphologist Walther Penck and academic adviser to Karl Haushofer's son, Albrecht Haushofer (who was executed in 1945 for his part in the plot to assassinate Hitler). In 1966 the East German dissident artist Ralf Winkler adopted the name A.R. Penck under which to work.

In Britain, one of the most significant contributions was that by Herbertson (1905). Influenced, like Vidal de la Blache, by the organismic analogy – he (p. 302) proclaimed that "[w]hile we may not be able to dissect our . . . terrestrial macroorganism into the organs, tissues, and cells of the vital organism, we can find in this idea a useful hint" – and seeing Geography as "the science of distributions" (p. 300), Herbertson argued that long-standing practices of studying the globe's geography according to political divisions was an unsystematic approach because such political regionalization reflected "comparatively unstable . . . human conditions" (p. 306). Rather, he contended, the key to a scientific Geography was to carve up the globe according to natural regions. In seeking to develop a system of regionalization, however, Herbertson (p. 300) faced two questions: 1) "What characteristics should be selected to distinguish one region from another?," given that "[s]ize is not a sufficient guide"; and 2) "How can we determine the different orders of natural regions?" The key, he maintained (p. 309), was that in any delineation of natural regions, such regions "should have a certain unity of configuration, climate, and vegetation." Although a region's "ideal boundaries" were the "dissociating ocean, the severing mass of mountains, or the inhospitable deserts," he recognized that the lack of such entities in much of the world meant that, generally, the boundary between one region and another "is not at all well marked, [such that] the characteristics of one region melt gradually into those of another" (p. 309). Nevertheless, for Herbertson natural regions were the basis for economic and cultural regionalizations. Thus, he held (p. 309) that recognition of natural regions would give "the historian a geographical foundation for his investigations into the development of human society, such as he has not hitherto consciously possessed."

Following Herbertson's outlining of a method to determine natural regions, Unstead (1916: 231) questioned what exactly was meant by this term, asking whether it was "intended to relate only to natural, in the sense of physical, conditions, or to indicate the regions which are marked out by natural, as distinct from artificial, boundaries." At the center of this questioning was a debate about the nature of Geography. Viewing regions as "geographical units" (p. 238), Unstead (p. 232) felt that regions defined according to physical conditions could be thought of as natural regions but they were "not geographical regions, for they take but little account of the living inhabitants of the Earth's surface." The key to developing an understanding of "geographical regions," he postulated (p. 234–35), was

> not to bring together on separate maps certain analyzed factors, such as relief, structure, temperature, rainfall, natural vegetation, etc., and to compare these elements, but to take their combined effects as

they work themselves out, i.e. to take the actual complex of physical and human conditions, to regard this as a closely interrelated whole, to observe the predominant characteristics of this complex in different parts, and so, by a synthetic rather than an analytic method, to arrive at the determination of regions with common characteristics.

Thus, for instance, what was important was to consider "the effects rather than the elements of climate" (p. 234). Such an approach meant that economic geographers and demographers would play central roles in compiling geographical regions.

In addition to moving the debate from that of "natural regions" to "geographical regions," with the latter incorporating to a much greater degree than previously human activities (even if environmentally directed), Unstead's intervention was important for three reasons. First, despite how Regional Geography has often been characterized in recent decades as Other to Quantitative Geography for its alleged focus on description of unique areas of landscape (e.g., Burton 1963; Hart 1982), Unstead argued that statistical analysis was an important element of Regional Geography – "To a considerable extent," he maintained, "the characteristics of the regions are capable of quantitative expression, and should be so expressed wherever it is possible" (p. 238). Second, by incorporating to a much greater degree the human element in regional-izations, Unstead (p. 241) appreciated that, because human activities change over time, geographical regions would "have to be reconsidered and reconstituted at intervals." Regions, in other words, should not be seen as temporally and spatially fixed but as dynamic. Finally, whereas Herbertson had constructed a top–down regional hierarchy by starting his analysis at the global scale and dividing the world into smaller pieces, Unstead contended that regionalization should proceed bottom–up "from small districts to large regions" (p. 235), with such "small uniform districts . . . of the first order . . . being combined with others adjoining them to obtain larger areas . . . which[,] though necessarily less uniform[,] would yet have many features in common" (p. 238).

Other geographers explored regionalization, if from a slightly differ-ent perspective. Fleure (1919) argued for a focus on what he called "Human Regions," with regionalization based upon human achievements (or the lack thereof), and determined seven such regions: of hunger; of debilitation; of increment; of effort; of lasting difficulty; of wandering; and industrialized. Such a focus, he averred (p. 94), avoided approaches which used "only the regions based upon relief, climate, and vegetation" and which were "apt to treat man too much as a creature of circumstance, as fitting himself into his natural surroundings." However, unlike natural regions' boundaries, which he saw as capable of providing fairly distinct and sharp divergences between geographically proximate locations, Fleure suggested (p. 105) that "the bounds of a human region must be conceived as zones, not as lines." For him (p. 105), human regions' boundaries were not intended so much to be understood as geographical walls by which landscapes are partitioned but are to be viewed, rather, as "intermediate zones" which "play the part of assimilators; they fuse diverse contributions to civilisation, and if they are sufficiently large they impart to the fusion a distinctive personality." Whereas Herbertson had maintained that regions must be based upon Nature, Fleure asserted (p. 94) that exploring both natural and human regions would highlight "more vividly the continuous interaction and interpenetration of man and environment and the cumulative alteration of both man and the earth with the unfolding of history."

In Germany, the regional tradition was perhaps most developed at the end of the nineteenth century by Alfred Hettner, who saw Geography as a chorological science and drew upon notions of Landschaft to develop regionalizations of the Earth's surface (Box 4.2). In this tradition, regions (raüme) were typically approached as layered entities, starting quite literally with an area's bedrock, surface morphology, climate, and drainage, and then moving up through a series of layers involving the distribution of plants and animals, to human settlement, economy, and population (Elkins 1989). For Hettner, however, focusing upon regions was a way of overcoming the Nature/Culture dualism extant within much German Geography of the period. For him, regions were the central analytical constructs of chorological Geography, though he believed it important to differentiate between, on the one hand, completely formal inter-relationships of similarity and difference and, on the other, functional relationships (Harvey and Wardenga 1998: 135). Moreover, because for Hettner geographic facts did not exist in and of themselves but were the results of chorological observations (in other words, "facts" were not independent of the process of observation), "regions were not ideals to be 'found' in reality, but were 'produced' through a methodologically

---

Box 4.2 LANDSCHAFT

The term *Landschaft* has caused some confusion outside German Geography, because of its etymology. According to Martin (2005: 176–77; see also James 1934), the term originally referred to an extent of territory more or less uniform in nature. However, after the fifteenth century it came to be used by German artists to refer to visual impressions of the landscape, though without any connotations of areal extent. In German Geography in the early twentieth century, *Landschaft* was used to connote a small region, as the term seemed more appropriate than the word for a larger region (*Gebiet*). Combining physical and human geography, if in a somewhat environmentally determinist manner, the term *Landschaft* came to have holistic connotations. Thus, a "*Landschaft* represented more than the sum of its parts [and] was [considered to be] a real and given object existing in its own right, and studied as such" (Arntz 1999: 298). Geography itself was widely viewed as *Landschaftskunde* ("landscape science").

---

systematic process of regionalization" (Harvey and Wardenga 2006: 428). This was a position in contrast to Kant's view of Geography, in which pure reason and logic allow the observer to understand the world through reactive contemplation largely divorced from engagement with that world. In Hettner's opinion, regions were real things and the ultimate objective in the chorological viewpoint should be "the recognition of the characteristics of countries and localities, growing out of the interrelationships of different natural realms and their various manifestations, and the consideration of the whole earth's surface according to its natural divisions into continents, regions, natural provinces, and localities" (quoted in Van Cleef 1930: 356).

Ideas concerning regionalization were also explored by the political geographer Friedrich Ratzel (Mercier 1995) and the geomorphologist Albrecht Penck. For his part, Ratzel (1897) saw regions as cultural/political units but viewed them in naturalistic terms as biological organisms intent on expanding into areas of "weaker peoples" (see Chapter 5 for more on Ratzel). Penck (1927), on the other hand, argued that regions and organisms are quite different – whereas he saw the latter as essentially indivisible, the former could be chopped into smaller units or amalgamated into larger ones. Moreover, for him Geography's focus should not be on individual homogeneous areas (what he called "chores" (Box 4.3)) for the purpose simply of categorizing them but, rather, on how such chores fit together to form larger territorial units (the Landesgestalt), as the landscape's structure can "only be fully understood and appreciated by studying the manner in which its individual regions are grouped, by considering what patterns they form, and by endeavoring to comprehend its geographical configuration" (p. 640). Like several other German geographers, Penck adopted the terminology of music in his conceptualization, suggesting that there is a harmony and rhythm which flows through the landscape, and even that the Landesgestalt formed by myriad chores could be characterized as a "symphony," with each chore playing its part in the grander work. (Interestingly, Howitt (1998) has suggested that the language of music may have relevance for understanding geographical scales, though in a different manner (see Chapter 1).) Finally, whereas many contemporaries believed Nature to be the landscape's primary shaper, for Penck it was humans who were the greatest single factor in determining its character.

Regional Geography in the US likewise largely initially focused on so-called natural regions, upon which economic or other humanly produced regions were then superimposed (e.g., Fenneman 1914). Indeed, according to Joerg (1914: 55–56), given that the "demand that [regions] be as homogeneous as possible . . . [was] rarely fulfilled by artificial units, such as political divisions, . . . the use of natural regions as the fundamental units of geographical investigation" was generally considered to be the only true basis for Geography as a science (see also Dryer 1912: 74, who argued that "[t]here is no thoroughfare for scientific regional geography by the way of political units and the necessity for basing it upon natural units is imperative"). Joerg (1914: 58) suggested there were two methods by which natural regions could be delineated: the inductive (which "first visualizes the individual regions and then groups them together into larger units") and the deductive (which "starts with the larger unit and subdivides it into its various parts"). By the end of World War I, however, the rise of a group of scholars who had trained as geographers (rather than as geologists), many of whose overseas wartime experiences had led them to be interested in differences in cultural and economic landscapes, began to bring about a move away from efforts to delineate natural regions (James 1954). Nevertheless, much of this latter work adopted a Ratzelian and/or environmental

---

Box 4.3 CHORES

Hartshorne (1939: 440) notes that the term *chore* was introduced by German geographer Johann Sölch to refer to an area of any size that was *relatively* homogeneous in all its geographical factors. Consequently, any *chore* can be divided into smaller *chores*, with each division resulting in a higher degree of homogeneity until a perfectly homogeneous unit is attained. According to Hartshorne, however, Penck used it in a slightly different sense, wherein a *chore* was seen to be *the* smallest possible land unit ("indivisible cells, so to speak"), various numbers of which would be added together to form larger entities. For his part, Hartshorne did "not follow this usage, not only because it changes the meaning of a term as the inventor defined it, but also because there can be no smallest units . . . [, for we] may continue the process of division indefinitely and our subdivisions are no less (and no more) real units than those we divided." Hartshorne's approach, in which any absolute space can be further divided almost infinitely, has the potential to create myriad chaotic conceptions (Sayer 1984) by breaking connections between objects which exist in the landscape but which are necessarily connected together. Penck's approach, on the other hand, allowed for a certain minimum beyond which absolute spaces cannot be subdivided, which provides for a greater likelihood that necessary relations between objects would be maintained.

---

determinist approach, with economic and political regions largely viewed as products of Nature – for instance, American Geographical Society President Isaiah Bowman (1924) saw Nature as the determinant of the human regionalization of South America and even went so far as to argue that the Andes' ragged physical geography could cause revolution (N. Smith 2003: 71–76), whereas Dryer (1915: 125) conceptualized regions as spatial units in which layers of humanly created phenomena sit atop underlying layers of natural phenomena, such that "[e]conomic function . . . furnishes . . . a guide for the delimitation and classification of natural regions."

In sum, with the exception of the Vidalian tradition, for the most part early conceptualizations of the region either assumed that Nature was the only basis upon which objective regions could be based or, in the case of regions based upon human activity, were riven with environmental determinist thinking. Furthermore, many writers saw regions in organismic terms, an approach consonant with the then-extant ecological tradition in Urban Geography (see Chapter 3). However, as the discipline began to move toward environmental possibilism, so did arguments about regions and the process of regionalization. For instance, though he saw regions in organismic terms, Unstead (1916: 240) declared that "Nature makes possible a number of developments, but which one will appear depends upon the incalculable human factors." Meanwhile, Tower (1908: 526) contended that regionalizations which adhered "to the strictly physiographic bases of classification" were too limited because they ignored humans and failed to see that animal and human life might respond differently to different configurations of natural environments. Rather, he averred (p. 526), whilst natural physiographic regions could be seen "as supplying a basis for certain general relations," it was necessary to "add new bases of division," such as the local particularities of climate, so as to "permit a more detailed grouping of human responses" to regional environmental conditions. Only through such a schema, he declared, could Geography "attain its highest development and become, in fact as well as in definition, the study of the earth in its relations to life" (p. 530).

Outside Geography, the US sociologist Rupert Vance (1929) argued for a position which saw the environment not as determining regionalization but as providing the context within which humans differentially adapt to it, in the process creating regions. Moreover, he accepted that human activity could have significant impacts upon the regionalizations given by Nature. Thus, he maintained that "[e]conomic factors, such as values per unit of weight and distances to markets, may determine the extent and distribution of plant production" (p. 214), with the result that "the region becomes the culture area characterized not only by common physical traits but by common culture traits." Consequently, although the Nature-given region may start off as "man's stage," eventually it "becomes . . . his handiwork and his heritage" (p. 215). For his part, Platt (1935) proffered that different types of processes lead to different types of regionalizations: natural processes (e.g., geologic) result in regions of static areal homogeneity, whereas human processes generate regions defined by their functional unity, with the result that the latter are changeable and heterogeneous.

THE REGION MID-CENTURY

Whereas in the early part of the twentieth century regions had generally been defined on the basis of Nature, by the 1930s growing numbers of geographers were beginning to question this approach. This was partly the result of the maturation of Human Geography, which had increasingly separated from its origins – particularly in the US – in Physical Geography and Geology, partly the result of the waning of environmental determinism, and partly a reaction to the economic crisis of the 1930s, wherein the rise of city and regional planning to deal with the Depression provided impetus for thinking in regional terms. For instance, although Renner (1935: 137) accepted as valid the assumption that "an area demarked by geomorphic, climatic, or edaphic factors . . . is a unit habitat impelling enough to produce unity in cultural affairs," he also suggested that "it has probably led to many incorrect conclusions in human economy." Hence, he wondered, could "regions be recognized and their boundaries drawn by approaching the matter from the non-physical side?" In other words, could "an area's 'regionality' be discovered and the region delineated through the measurement of its social data?" He based this questioning upon two postulates.

First, he averred that although natural regions may initially set out the parameters within which human activity takes place, such human activity has so transformed these regions that it now constitutes the "real 'regionality' of the area." Consequently, it made sense to define regions on the basis of human adaptations to the environment rather than Nature and to separate the concepts of humanly produced and natural regions, since different groups of humans could develop different cultural responses to the same set of physical conditions. Second, he contended that "the visible manifestation of these geographic adjustments is a cultural landscape or combination of landscapes whose outer limits form the boundaries of the region." Based upon these two postulates, Renner maintained that regions could be described solely through their cultural attributes and that these attributes could be discovered either through intensive fieldwork – which he believed preferable but costly and time-consuming – or through a statistical analysis of data designed to measure the "relationships underlying and giving rise to that cultural landscape" (p. 138). (Significantly, Renner's call for a statistical approach to region-alization matched a number of ad hoc efforts to use statistics to define regions which had been ongoing since the 1920s (e.g., Baker 1926

1927a, 1927b.) Nevertheless, despite its focus upon various human activities, such regionalization still tended to be based upon the similarity between places that were spatially proximate to one another, rather than upon any functional or processual bases – Woofter (1934), for instance, devised a schema wherein he allocated counties to agricultural sub-regions based upon how alike they were according to various indices, an approach which did not recognize that places which are dissimilar in terms of appearance may nevertheless be integrally interconnected by process (his regions were, then, chaotic conceptions).

US Geography at mid-century, however, was arguably most shaped by Richard Hartshorne and Carl Sauer (Box 4.4). For Hartshorne, the fundamental geographic unit was the region, an argument he spelled out in his 1939 opus *The Nature of Geography*. Adopting a vaguely Kantian approach to studying the Earth's surface, Hartshorne (1939: 461–62) believed there was need for a "chorological science" (Geography) that "interprets the realities of areal differentiation of the world as they are found, not only in terms of the differences in certain things from place to place, but also in terms of the total combination of phenomena in each place, different from those at every other place." The result was an approach which focused upon the idiographic, even as Hartshorne argued for a scientific way of so doing. However, Hartshorne's suggestion (p. 463) that, "in studying the interrelation of [geographical] phenomena, geography depends first and fundamentally on the comparison of maps depicting the areal expression of individual phenomena, or of interrelated phenomena," his approach was highly spatially fetishistic: the spatial distribution of one thing on the Earth's surface explained the spatial distribution of something else – space, in other words, explained itself. Furthermore, because, he argued (p. 464), there were "no set rules for determining which phenomena are, in general, of geographic significance," when generating regionalizations it was perfectly acceptable to engage in "the arbitrary device of ignoring variations within small unit-areas so that these finite areal units, each arbitrarily distorted into a homogeneous unit, may be studied in their relation to each other as parts of larger areas" (p. 465). The result was that the "determination of [geographical] divisions at any level involves . . . the subjective judgment as to which features are more, which less, important in determining similarities and dissimilarities, and in determining the relative closeness of regional interrelations" (p. 466).

Box 4.4 THE BERKELEY AND MIDWESTERN SCHOOLS OF GEOGRAPHY

In the 1930s, two "schools" of thought dominated US Geography: the so-called Berkeley School and the Midwestern School. The former was centered on Carl Sauer (1889–1975), who arrived at the University of California–Berkeley in 1923 and taught there until his retirement. The latter incorporated academics at a number of Midwestern universities, including Chicago and Northwestern, Wisconsin–Madison, Minnesota, Michigan and Michigan State, and Pennsylvania State University, although arguably the central figure was Richard Hartshorne (1899–1992), who taught initially at Minnesota (1924–40) and then at Madison (1945–70). Both Sauer and Hartshorne received their doctorates from the University of Chicago (Sauer in 1915, Hartshorne in 1924) and both drew on ideas from German Geography (especially Alfred Hettner), but they adopted quite different approaches toward Geography as a discipline. Sauer was particularly influenced as a student at Chicago by the geologist Rollin Salisbury and the plant ecologist H.C. Cowles. Once at Berkeley, he developed strong connections with Anthropology (whose two faculty members at the time had been students of Franz Boas, the so-called "Father of American Anthropology") and, later, the biological sciences. For his part, Hartshorne studied in Germany with Hettner. His brother was the philosopher Charles Hartshorne, who studied in Europe with Edmund Husserl and Martin Heidegger before teaching at the University of Chicago (1928–55), Emory University (1955–62), and the University of Texas at Austin (1962–76).

Sauer argued that Geography's principal focus should be what he called the "cultural landscape," specifically the process of its change over time. He argued that the cultural landscape – and particularly its material manifestations – could be read so as to give insight into the cultures which created it. This focus, Sauer hoped, would move Geography away from the environmental determinism that had previously predominated. For him (1931: 622), though, "Man" should not "himself [be] directly the object of geographic investigation." Rather, it was Culture, not humans, that made the cultural landscape: as he famously put it (1925: 46), "Culture is the agent, the natural area is the medium, the cultural landscape the result." This idea would

continued

Box 4.4 continued

ultimately result in the idea of culture as "superorganic," an idea pushed by Zelinsky (1973), who was one of Sauer's students and who saw Culture as a force that acted beyond, and somewhat independently of, humans (for critiques, see Duncan 1980; Mitchell 2000). Although Sauer accepted that regions ("culture areas") were important, he did so only in the context of what they said about landscape evolution – in other words, it was not the region itself which should be Geography's focus but what the region revealed about how humans transformed the environment as presented to them by Nature.

Hartshorne's *The Nature of Geography* was largely written as a response to the work of Sauer and his protégé John Leighly. In particular, Hartshorne did not see studying cultural differences as a sufficiently sound basis for defining Geography as a discipline. Instead, he viewed Geography's purpose primarily as observing a region's uniqueness and making the region understandable to others. Whereas Sauer concentrated upon process, Hartshorne believed Geography's focus should be upon pattern in the landscape – what he called "areal differentiation." This perspective is evident in Hartshorne's (1935a: 801) argument that "Geography is . . . fundamentally a science of places or areas rather than of supposed relations." Although he accepted that "in its study of places it is concerned with the causal relations that may be found between the different elements that go to make up the landscape, or the 'character,' of the place or area," he nevertheless contended that "[t]he study of the relationship . . . is subordinate to, and geographically significant only as a part of, the study of the area." Hartshorne (1939: 195) sought to avoid the question of whether regions really exist in the world by suggesting that geographers could never "find objective means for observing them." Consequently, the most geographers could say is that "any particular unit of land has significant relations with all the neighboring units and that in certain respects it may be more closely related with a particular group of units than with others, but not necessarily in all respects" (p. 275). Hartshorne's disdain for process resulted largely from his Kantian philosophical background, which saw the study of change over time as being the domain of History whereas difference over space was that of Geography, and his fear

continued

Box 4.4 continued

that a focus upon causal relationships between society and Nature would resurrect the discredited environmental determinism of the early twentieth century. Although Sauer was a fierce critic of environmental determinism, he nevertheless rejected Hartshorne's argument that Geography should avoid speculating about landscape change over time or seeking to delineate causal relationships between humans and their environment, calling this the period of Geography's "Great Retreat" (Sauer 1941: 2).

Both Sauer's and Hartshorne's influence on the discipline was made by their writings, but also in other ways. Although Sauer was the advisor to more Ph.D. students than was Hartshorne (over 40 at Berkeley (Mitchell 2000: 21) compared to Hartshorne's ten at Wisconsin (Martin 1994: 489)), some 65 to 70 geographers from more than 20 universities worked under Hartshorne during World War II at the Geography Division of the Research and Analysis Branch of the Office of Strategic Services, a fact which gave him great influence in shaping early post-war US geography.

Although Hartshorne saw the region as a fundamental geographical unit, he nevertheless proclaimed (p. 275) that geographers had "not yet discovered and established regions as real entities, [and that] we have no reason ever to expect to do so." Whereas he recognized that "there are entities in reality that are not 'naïvely given,' but [that] must be discovered by research" and that "if [regions] can be discovered, they can be established definitely, within a reasonable margin of certainty," Hartshorne also asserted that "[w]e have not achieved that in one single case" and that "establishing the boundaries of a geographic region . . . presents a problem for which we have no reason to even hope for an objective solution" (p. 275). In contrast to those who argued that regions existed as material things, in Hartshorne's view "[t]he regional entities which [analysts] construct . . . are therefore in the full sense mental constructions; they are entities only in our thoughts, even though we find them to be constructions that provide some sort of intelligent basis for organizing our knowledge of reality" (p. 275) – a classic Kantian position. Consequently, the Earth's regionalization was to be based not upon its materiality but upon logic – "a task that involves a complete division of the world in a logical system, or systems, of division and subdivision, down to, ultimately, the approximately homogeneous units of areas" (p. 465). Accordingly, parroting Hettner, Hartshorne (p. 290) argued that "one cannot speak of true and false regional divisions, but only of purposeful and non-purposeful" ones, with such purposefulness determined by the individual researcher.

Hartshorne's conceptualization both of the region and of the nature of Geography as a discipline had several significant implications. First, according to Hartshorne the fact that "the world . . . is not divided into distinct areal parts" meant that "the fundamental function of geography – the understanding of the differences between different areas – requires the geographer to divide the world arbitrarily into areal parts" (p. 361). The result of such a perspective is that Hartshorne saw the region, arguably Geography's central concept, as no more than a figment of the imagination. Second, and relatedly, because regions did not inhere to the world (p. 362), Hartshorne argued they could "be divided arbitrarily into parts that, at any level of division, are, like the temporal parts of history, unique in total character" (pp. 467–68). Likewise, they could be combined in any manner of fashion, for only the Earth itself was Geography's "individual, unitary, concrete object of study" (p. 262). Hartshorne's view of space was one which saw it in absolute rather than relational terms, as something that could be chopped up into pieces or amalgamated into larger segments with little attention paid to the natural and social processes which might link various parts of the landscape together. Third, Hartshorne rejected the argument that Geography "is the study of the landscape, or of land-scapes" (p. 159), suggesting that the term "landscape" had been used within Geography simply as a rather imprecise synonym for "region" (p. 160), with writers using it "to mean something about area without necessarily implying any limit to the extent of area" (p. 161). "Landscape" as a concept should therefore be rejected, he averred. Moreover, for Hartshorne natural and cultural landscapes qua regions were not organic material entities, the active production of which by humans driven by broader economic and political forces it was Geography's job to understand. Rather, a landscape was instead simply a "superficial phenomenon" (p. 165) to be described for purposes of cataloguing every conceivable idiosyncrasy it might contain, with the "discovery, analysis and synthesis of the unique . . . represent[ing] an essential function of science" (p. 468). Consequently, although Hartshorne recognized that the same elements might be found in different regions, he maintained that "the combination of all of them in the actual mixture as actually found, occurs but once on the earth" (p. 393).

If Hartshorne represented one of the key writers on regions in mid-twentieth century US Geography, the other was Carl Sauer. Although the two are often seen to be at polar extremes with regard to their view of Geography, in fact they shared similar perspectives in many regards. For example, both were heavily influenced by the German geographic tradition and both agreed that Geography should focus upon areal differentiation, even to the point where Hartshorne (p. 237) cited Sauer as a fellow believer in this respect (see Sauer 1941: 6 for the latter's view). Equally, both saw Geography as a synthetic discipline, both rejected grand social theory (largely out of a concern as to how it had been used during the era of environmental determinism), and both saw descriptive mapping as the basis for making scientific claims about the world. However, Sauer did not elevate the region to the vaunted status that did Hartshorne, largely because he was little interested in regionalization – as he (1924: 22) put it, "[r]ecognition of characteristics, not the tracing of boundaries, is the urgent matter in the study of the natural region." Indeed, Sauer maintained (p. 22), it was "rather im-material whether a field of study is coextensive with a natural region, is a fragment thereof, or cuts across natural boundaries." Unlike Hartshorne, who was largely focused on understanding contemporary patterns of phenomena across the landscape and who argued (1935b: 947–48) that, "in contrast to the historian, the geographer is not concerned primarily with the sequence of events" but is (p. 956) "only indirectly interested in the *processes* that led to [a landscape's] historical development," Sauer was particularly interested in landscapes' historical change. Consequently, following German geographer Gustav Braun, he contended that Regional Geography's proper focus should be upon how cultural landscapes (*Kulturlandschaft*) arise out of the natural, pre-given landscape (*Naturland-schaft*) (Sauer 1924: 24). Although he conceived of Geography as "culture history in its regional articulation" (Livingstone 1992: 297), Sauer (1964/1987: 155) declared that he was interested primarily in "what made the life of a people in a predefined area . . . significant and characteristic," rather than in how regions are delineated. Thus, he took "the cultural landscape" to be "a particular land as it appears in the occupation and expression of the people who are in it," a land wherein

"dimensions [are] not considered [and] boundaries [are] not involved."
As a result, he averred, "I didn't have to worry about boundaries."

Although Sauer's interest in landscape was anathema to Hartshorne, Sauer did not quid pro quo reject Hartshorne's central concept, the region. Indeed, in his Association of American Geographers Presidential Address he (1941) counseled geographers to become regional specialists. Nonetheless, if Sauer and Hartshorne had varied ideas about what should be the focus of geographical study and the need for stringent regionalization, it was equally the case that they had different ideas about what was a region. Whereas for Hartshorne the region was essentially an arbitrary mental construction, for Sauer regions were real, material things which could be determined objectively in the landscape. Hence, he suggested (1925: 22–25),

> the phenomena that make up an area are not simply assorted but are associated, or interdependent. To discover this areal 'connection of the phenomena and their order' is a scientific task ... [and this task fails] only if the non-reality of area be shown ... The objects which exist together in the landscape exist in interrelation. We assert that they constitute a reality as a whole that is not expressed by a consideration of the constituent parts separately, that area has form, structure and function, and hence position in a system, and that it is subject to development, change and completion.

As a result, Sauer (p. 26) considered the landscape "in a sense" to have "an organic quality." Cultural Geography, he averred, should proceed "from a description of the features of the earth's surface by an analysis of their genesis to a comparative classification of regions" (Sauer 1931: 622).

There are four important elements to Sauer's view of landscape which have bearing upon his implicit conceptualization of the region. First, although he saw in the concept of landscape the potential to link morphology with process, he tended to emphasize the former over the latter, focusing less on processes of cultural change than on the chorological expression of the material products of such change (Solot 1986: 512, 516). Put another way, how various areas qua regions came to contain particular geographical phenomena was of secondary importance to what they contained and how what they contained interacted as part of an organic entity. Second, for Sauer the making of landscapes was a rather individualistic and voluntaristic process, rather than a collective one in which actors are caught up in dialectical relationships with various social and natural structures. Third, despite his efforts to use the concept of landscape to "bridge the gulf between space and society [and thus] diffuse the dualism engendered by the absolute conception of space" in which "[e]vents, objects and processes do not constitute space but happen 'in space'" (Smith 1989b: 108, 104), in fact Sauer actually separated ontologically the landscape's spatial form from the people who produce that form. In this respect, his previously quoted comment (Sauer 1964/ 1987: 155, emphasis added) that the cultural landscape represents "a particular land as it appears in the occupation and expression of the people who are in it" rather than, say, who *make* it, is telling. Finally, because he argued (Sauer 1931: 623) that the landscape is a material entity, the study of which should focus upon its "visible, areally extensive and expressive features," Sauer's was ultimately a positivist approach to understanding landscape and area (and, hence, regions), for it viewed as valid only that knowledge about the landscape which came from analysis of those objects which are (or potentially are) directly observable.

Whereas Hartshorne and Sauer were, perhaps, the two most central players in mid-century US Geography to write about regions, if in different ways, others were also involved in these debates, especially the debate over chorography. Peattie (1935: 172), for instance, anticipating future challenges to the chorographic tradition, suggested that whilst "[g]eneralization . . . is very difficult," too often had chorography been thought of as "an end in itself, the goal to be sought." There was, he averred, "too much chorographic counting of cabbages without elucidation of principles." For him, too many geographers were "simply landscapists," and although geographers should "start with the cultural landscape," they should "not stop there." Likewise, Dodge (1935: 174) argued that rather than focusing simply upon any region "that the individual geographer has an opportunity to approach," Geography should develop a more scientific way of identifying and analyzing regions. Meanwhile, Whittlesey (1954) saw the region as "an area of any size [that is] homogeneous in terms of the criteria by which it is defined [and] throughout which accordant areal relationship between phenomena exists" (pp. 21–22). For him, regions are real entities possessed of "a quality of cohesion that is derived from the accordant relationship of associated features," whilst the "observation and measurement . . . and search for . . . areal relationships among . . . these phenomena [constitute] the regional method or the procedure for discovering order in earth-space" (p. 22). Such order, he proclaimed (p. 22), "is expressed in the form of regional patterns made up of specifically defined characteristics and distributed within clearly outlined borders."

Although issues of regionalization and Regional Geography were perhaps less central in the UK than they were across the Atlantic, writing about economic, political, and cultural regions was also common in the post-World War I era and followed, in many ways, developments in the US. Thus, one multi-author project, based upon the belief that the "purpose of regional geography is to describe the regions of a country as they are and to discover the causes that have made them what they are" (Ogilvie 1928/1952: 1), generated essays on 24 regions of Britain, with the natural environment presented as the shaper of human regions – a perspective that drew on the idea of natural regions, which continued to dominate British Geography in the 1920s (e.g., Roxby 1926). In the 1930s, geographers such as L. Dudley Stamp engaged in detailed regional analyses to determine how the Depression was playing out across the country (Clout 2003b) and many geographers worked with, and on, local and regional planning departments tasked with addressing unemployment's effects. This tradition carried over into the World War I era, when myriad geographers were seconded to write regional monographs of various parts of the world (Bennett and Wilson 2003; Unwin 2006). These developments mirrored events in the US, where many geographers were employed during World War II to write descriptive regional geographies of areas that were of strategic interest and continued doing so after the war's end (Butzer 1989). Stamp and others also engaged in regional studies of Continental Europe, as well as various parts of the British Empire (e.g., Steel and Fisher 1956), with such work generally involving constructing detailed studies of land use, house types, economic activity, and the like, and resulting in their writing brief monographs. For his part, the historical geographer H. Clifford Darby completed a major regional study of the Fens of East Anglia in 1940 (Darby 1940a, 1940b) which drew upon the Vidalian tradition (Clout 2003b). Nevertheless, although regional description continued to dominate British Geography in the early post-war period, criticism of using Nature as the basis of regionalization began to grow louder in Britain in much the way that it had in the US – arguments initially made by geographers such as Unstead in the 1910s were reinforced, for instance, by Dickinson (1939), who argued that regions are human creations, and Stevens (1939), who suggested that not even a Nature-given coalfield had unity if it were divided between two nations (for an Anglo-Canadian perspective, see Kimble 1951).

In France, meanwhile, the Vidalian region, with its catholic sense of *milieu* and *genre de vie*, gradually gave way to a focus upon areas of consumption and production within the context of massive urbanization. This was particularly associated with material changes in French life, especially demographic renewal in the 1940s, which raised questions of how economic and urban development could be both stimulated and managed (Claval and Thompson 1998). Consequently, the long-standing interest in rural France which had dominated Regional Geography began to be replaced by a focus upon the Regional Geography of resource endowment and how particular urban areas could be developed as growth poles at the center of urban regional networks, although there was recognition that "the intimate knowledge of these spatial combinations which are the *paysages* is indispensable to those who wish to evaluate a territory's [economic] potential [and that it] is impossible to separate even a clearly individualized *paysage* from the neighboring *paysages* with which it maintains functional relations [*rapports de complémentarité*]" (Juillard 1962: 490–91; Herod translation). Similar developments took place in Germany. Thus, although German Geography had long been dominated by a Ratzelian environmental determinism that manifested itself in an organicist view of regions, beginning in the 1930s a more functional approach began to emerge. This diverged from the traditional attention to "civilization" and landscape and focused, instead, upon the region as an economic entity. It was particularly associated with Christaller (see Chapter 3), and would result, by the 1960s, in a concentration upon regional spatial organization embedded within a systems theory approach (Hoekveld 1990).

REGIONS AND SPATIAL SCIENCE

By the 1950s, how regions were theorized had begun to change. Disciplinary histories have generally presented this period as one of intense intellectual restructuring, as the "naïve description" of Hartshornian Regional Geography was increasingly replaced by the quantitative revolution's "systematic approaches," with Schaefer's (1953) paper critiquing Hartshorne and his ilk for conceptualizing Geography as an integrative field focusing upon the idiographic usually seen to be the opening salvo in this regard (Box 4.5). However, whereas Schaefer's attack is often characterized as a call for developing a Systematic Geography in opposition to Regional Geography, a more careful reading shows that such a dualism is perhaps not as stark as commonly asserted. Hence, at the beginning of the twentieth century Herbertson (1905: 311) had seen Regional Geography as capable of generalizing, advising that "it is desirable that we should find out what regions can be grouped together, so that by studying, say, fifteen or twenty types, the main features of the greater part of the world are learned, and merely the details of the varieties of each type have to be mastered later on. In practice this results in a saving of much time, and adds greatly to the value of geography as an intellectual exercise." Hartshorne (1935b: 954) himself had even argued that "regional geographers have always had to depend . . . on . . . 'systematic geography' . . . in which any particular feature or group of features are studied comparatively in different areas of the world." Meanwhile, Schaefer (1953: 230) declared that "regional and systematic geography are codign [sic], inseparable, and equally indispensable aspects of the field." For his part, Campbell (1994) has claimed that Hartshorne has been unfairly tagged as engaging in simple description when, in fact, he was a methodological generalist.

Equally, although the 1950s and 1960s saw the emergence of what is often called the Spatial Science "paradigm," in fact regional monographs continued to be written. In France, although Regional Geography became less common during this time, in the 1970s the growing interest in the rural past saw something of a revival of the Vidalian approach (Claval 1984; Claval and Thompson 1998). Similar trends were evident in Germany where, although an anti-regionalism was evident in the 1950s, Regional Geography enjoyed a resurgence in the 1970s, the result of the fact that 1960s quantification did not reach as deeply into German Geography as it did in the US and Britain and also that German Geography's disciplinary structure has generally encouraged fieldwork outside Germany (Lichtenberger 1979; Hoekveld 1990). In Britain, some geographers continued to write regional studies redolent of the descriptive approaches of the pre-war era, with Dickinson (1976: 382–83) arguing that "[t]he true geographer is a regionalist [who] searches for interpretations of regional entities at any scale . . . as a terminus ad quem [a goal to be achieved], not an origo a qua [point of departure]." Likewise, in the US Hart's (1982) defense of Hartshorne and his call for a return

Box 4.5 THE SCHAEFER–HARTSHORNE DEBATE

In 1953, Fred Schaefer (1904–53), a professor of Geography at the University of Iowa who had been born in Berlin but had fled the Nazis, published a blistering critique of Hartshorne's approach to Geography. Primarily, Schaefer argued that Hartshorne had erroneously attempted to develop for Geography an "exceptionalist" identity, by which he meant that Hartshorne had argued that, because of its focus of inquiry (the idiographic), Geography was a discipline that required a different methodology from others. By way of contrast, Schaefer argued that although Geography's empirical focus may be different from other disciplines, this did not stop Geographers from adopting a methodology no different from other sciences. Indeed, he bluntly argued (1953: 231) that it was "absurd to maintain that geographers are distinguished among the scientists through the integration of heterogeneous phenomena which they achieve" and that there was "nothing extraordinary about geography in that respect." According to Schaefer, Geography should be a law-finding and law-developing discipline, even if he recognized that the laws it would produce were morphological (concerned with spatial form) rather than concerned with process, as was the case in other "mature" sciences. It was through finding and developing such laws on location that Geography could scientifically differentiate the Earth's surface into regions.

In response, Hartshorne wrote *Perspective on The Nature of Geography* as a means to defend his position, arguing that Schaefer had misunderstood him, that the proper focus for Geography was the detailed description of particular places, and that his approach was indeed a scientific one because it saw Geography as "concerned to provide accurate, orderly, and rational description and interpretation of the variable character of the earth surface" (1959: 21). As it happened, Schaefer had died of a heart attack before his initial article was published, and so was unable to respond to Hartshorne's attack of his critique. Nevertheless, Schaefer's paper served as a rallying point for a new generation of geographers and is frequently viewed as the catalyst for the so-called Quantitative Revolution in Geography.

to descriptive Regional Geography suggest that rather than a Kuhnian (1962) "paradigm shift" taking place in the 1950s, it might be more appropriate to imagine the discipline as having braided like a stream, with some branches expanding and others withering but not entirely drying up.

Generally, spatial scientists viewed regions in two ways. First, regions were often seen as a convenient aide to classification – the region was, Haggett (1966: 241) declared, "one of the most logical and satisfactory ways of organizing geographical information," whereas for Bunge (1962: 195) "Regional geography classifies locations and theoretical geography predicts them." In this regard, their approach was quite Kantian and not so different from what Hartshorne had argued – regions served simply as useful spatial devices for imposing order on the world's complexity. Consequently, much time and effort was spent developing taxonomic schemas by which to categorize various regional types for purposes of more clearly organizing geographical information. For instance, Bunge (1962: 26), who argued that the Spatial Science approach represented "the overt methodological abandonment of the concept of uniqueness," identified four regional types: uniform regions (the "classic geographic regions . . . that . . . can be fruitfully viewed as areal classes in a classification system" (p. 14)); experimental regions (regions created as part of a theoretical framework designed to understand the landscape, such as Christaller's hexagonal regionalization); nodal regions (as when a central place sits in the hub of an area); and applied regions (what Bunge saw as an effort to incorporate classificatory and theoretical approaches to "the real world"). Although he did not deny that certain types of regions "do seem to exist as concrete unit objects" (p. 25), for Bunge "the region" was essentially a unit that is spatially homologous to that of "class" in other fields, serving as a way to classify spatial data. Meanwhile, Haggett (1966) agreed with Bunge that regions represented means for general classification but also averred that there remained the issue of their absolute location, suggesting (p. 243) that "[h]owever they are classed, regions retain their locational uniqueness" – no two regions are in the same absolute location on Earth.

Second, many spatial scientists viewed regions as material entities which could be defined through the application of rigorous statistical analysis. Moreover, whereas the Regional Geography of the pre-war era had tended to center upon formal regions based upon Nature (a practice which Kimble [1951: 159] suggested involved "trying to put boundaries that do not exist around areas that do not matter"), spatial scientists largely focused upon functional regions, seeking to show *via* precise mathematical methods how flows of goods and people across the landscape created regions. For example, Zobler (1955: 84) contended that "simple inspection" of field observations was an insufficient manner by which to engage in regionalization and argued that statistical testing would "enable the investigator to validate [regional] boundary lines objectively by employing the concept of statistical significance, instead of relying on arbitrarily selected quantitative differences." Following from such arguments, Berry (1966, 1967) outlined a three-step process for identifying regions and their spatial limits: i) factor analysis/principal components analysis (as a basis for determining the bases of regional structure); ii) dimensional analysis (as a means to evaluate how similar are each pair of places analyzed with regard to their economic structure); and iii) grouping analysis (as a way to group contiguous sets of similar places into regions). Grigg (1967: 494), meanwhile, contended that regions are models of the world and that the growth of computing capacity in the 1960s allowed the greater use of analytical statistics (especially factor analysis), rather than simple descriptive statistics, to delineate them, all whilst recognizing that a "system of regions is established only as a first step in a geographical inquiry." Nevertheless, despite such efforts, spatial scientists did sometimes employ a somewhat vague definition of the region – Haggett (1972/1975: 6), for instance, argued that a region "is any tract of the earth's surface with characteristics, either natural or man-made, that make it different from areas that surround it." At the same time, Harvey (1969: 73), in his pre-Marxist persona, argued that whereas the concept of Geography as a chorological science of the idiographic rested on an assumption that space is absolute and exists independently of the social objects contained in it, the rise of nomothetic approaches to regionalization drew more on relativistic views of space that saw regions as "not unique or, at best, unique only within a selected coordinate system."

Decrying what he saw as Hartshornian Regional Geography's subjectivism, McDonald (1966: 516, emphasis added) maintained that regional study's goal,

> *particularly as geography takes a more objective view of the world* and attempts to sharpen this objectivity by increasing the accuracy of its measurements and broadening the base of its critical evaluations,

would seem to be to differentiate spatial units as regions on the basis of distributional aspects of a large part or the entire body of influencing factors acting on these units.

However, he submitted that there were two different approaches to the question of determining regional limits. The first, which dominated the first half of the twentieth century, defined regions on the basis of those of their characteristics which are susceptible to visual identification. However, because this approach generally lacked a mechanism for precisely circumscribing them, often it actually delineated regions on the basis of either a pre-existing boundary (such as a political frontier) or a particularly sharp boundary (such as a change in geology). On the other hand, the second approach, largely associated with the emergent Spatial Science, sought to define regions in terms of a variety of factors "which bear on [their] development, organization, and uniqueness" (p. 518) and to reveal quite hard-and-fast spatial limits to them. For McDonald this latter opened a Pandora's box of methodological issues because determining such geographical limits required having very precise spatial data. Moreover, regional delineation was often driven more by the availability of data than it was a process in which appropriate data were used to reveal a region's spatial extent, a practice which led to the determination of regions that were just as arbitrary as were Hartshornian ones. In response, McDonald advocated (p. 520) for a sort of middle ground "between observational approximation and the precision implied by mathematical analysis," with a region defined as "something more than a visual impression, yet something less than a statistical synthesis of the myriad factors exerting influence in space and over time." This could be achieved, he opined (p. 520), through taking into consideration only those elements "which have a decisive bearing on the region's unique- ness, as opposed to those, equally present or absent, whose presence or absence has no explicative quality vis-à-vis the region." This raised the question of how to determine whether an element was significant or not.

Despite spatial scientists' efforts to distinguish their approach to the region from that of the chorologers, in fact there were some signifi- cant similarities between the two. Whereas some early twentieth-century regional geographers had viewed regions in organismic/naturalistic terms, so too did some spatial scientists – Lösch (1938: 78) declared the region "a system of *various* areas, an organism rather than just an organ," whereas Haggett suggested (1972/1975: 421) that city regions could perhaps be thought of in ecosystem terms since they, "[l]ike watersheds, . . . need a constant flow of energy to maintain themselves." Equally, Sack (1974) contended that chorology and Spatial Science are not really that different in some crucial ways and that they are more like two ends of a continuum of explanation than they are polar opposites – for instance, both have a commitment to empiricism and both focus upon objects' spatial relationship (what Sack (p. 439) called the "physical geometric relationships of facts"). Nir (1987) argued that the division between Regional Geography and Spatial Science is not as great as has historically been considered because the latter, in its critique of the former, has really misunderstood the term "uniqueness" by equating it with "exceptional" and, from that misunderstanding, has built something of a strawman argument. Hence, he suggested (p. 188) that "unique phenomena, in spite of being unique, can be dealt with as part of a larger whole and can produce lessons and conclusions of general value, if only we understand that their uniqueness is relative." Consequently, using the analogy of humans – each human is unique yet shares commonalities with others – Nir contended that the region should be considered as a *holon* (from the Greek *holos*, meaning complete/whole) wherein, "inwardly, within the content of a certain framework, it constitutes a whole and something final for its components" yet "outwardly, the region is only one of the components which constitute greater wholes, such as states and nations" (p. 189). The result, he averred, was that "[j]ust as man – a unique entity – can be agglomerated to different groups, the same can be said of regions, also unique entities, which are composed of measureable [sic] or appreciable elements. Therefore the argument that the study of a region, being a unique phenomenon, cannot teach lessons applicable to other regions . . . is invalid." For their part, Symanski and Newman (1973: 352) argued that whereas nodal regions have typically been thought of as being about process whilst formal regions have been thought of as being about pattern, such that the former are treated as dynamic and the latter static, in fact this "dichotomy evaporates when it is realized that both classes of regions have a process and a pattern component, and that, in the end, all is process."

REGIONAL GEOGRAPHY RECONSTRUCTED?

The waning of Spatial Science's dominance in the 1970s led to a renewed interest in regions within Anglophonic Geography in the 1980s.

Interestingly, this renewed interest came from two quite different groups of scholars. First, there were a number who argued that Geography should refocus attention on regions and description of landscapes as part of a "back to basics" (Lewis 1985: 473) movement. In Britain, for instance, Watson (1983: 385) argued that "Geography must be concerned with the unique place," whilst in the US Hart (1982: 1) maintained that "[t]he idea of the region provides the essential unifying theme that integrates the diverse subdisciplines of geography" and that "[t]he highest form of the geographer's art is the production of evocative descriptions that facilitate an understanding and an appreciation of regions." Regional synthesis, Hart opined, was an essential twin to the "systematic approach" adopted in the 1950s, a twin that would save Geography from hobbling along "like a one-legged beggar" (p. 18). For his part, although he did not mention Regional Geography per se, Lewis (1985) presented a vision for the future that seemed to draw upon mid-century Regional Geography traditions when he suggested that human geographers would become better analysts by reinforcing their training in Physical Geography and History and, on a regular basis, "break[ing] loose from their library carrels, and light tables, and computer terminals and go[ing] outdoors" to simply observe "the immediate world that lies all about them" (pp. 472–73) – a position that not only smacked of conventional regional synthesis but also privileged local and regional scales of experience.

Pudup (1988), however, argued that because they failed to think critically about the process of description that was supposed to be at the heart of this approach, these efforts did little more than reintroduce a largely positivistic view of the region and how it was understood. Hence, for her, Lewis (1985: 374) presented a Regional Geography in which:

> [t]heoretically neutral observations are the basis for areal description[; t]he methodological questions are: where is it? what does it look like?[; i]nterpretations . . . are read off from these neutral descriptions[; n]o measure is taken of the epistemological basis of seeing[; and] regional synthetic accounts . . . use [information gathered about places] without regard for the ways it ha[s] *already* been socially constructed.

Likewise, Hart's epistemology incorporated a voluntarism and methodological individualism in which human agency is taken to be largely unconstrained, such that people are seen capable of making their histories and geographies pretty much as they choose, and in which there is a denial of the possibility that "the study of regions can be formalized . . . [and] that *a priori* theorization is a necessary part of geographical research" (p. 376). The result was a translation of "the traditional approach's defensive posture of the 1960s – 'we describe' – into an exclusively assertive posture – 'only we can describe'" and a concomitant "repositioning [of] traditional regional geography . . . to claim exclusively, among all geographical methodologies, the virtue of celebrating the unrestrained human will" (p. 376). Moreover, these approaches did not have a particularly sophisticated view of the region itself. Thus Hart (1982: 21–22) maintained there could be "no standard definition of a region, and . . . no universal rules for recognizing, delimiting, and describing regions." Consequently, he had a fairly spatially imprecise view of the region, believing that it "may be as small as a factory, a city lot, or a farmstead, or it may be the entire earth" (p. 23).

The second group who became increasingly interested in regions were Marxist geographers. As we saw in Chapter 3, in the 1970s many Marxist geographers (e.g., Harvey) and non-geographers (e.g., Castells) had sought to understand the urban's role within capitalism. However, within Marxist-inspired Geography the regional also became a point of interest. Such interest was largely a reflection of the intense economic restructuring then taking place in the industrial capitalist countries. For instance, Massey (1984) explored how economic restructuring in Britain was bringing with it new regionalizations and how these new economic landscapes were themselves impacting processes of restructuring as part of what Soja (1980) called a sociospatial dialectic. Likewise, in the US, Bluestone and Harrison (1982) examined how economic restructuring was heralding a dramatic restructuring of industrial regions. However, despite such interest, much of this work had a fairly unsophisticated view of the region and processes of regionalization. Thus, in her analysis, Massey (1984: 11) defined the region as "any sub-national area of any size," whilst Castells (1977) saw it as the space of production (in contrast to the urban, which he viewed as the space of consumption – see Chapter 3) but did not spend much time pondering how the region itself came about.

Moreover, within the Marxist camp there were two views of the region's theoretical status. These divergent understandings had their origin in debates over whether or not the production of capitalism's geography was integral to the accumulation process. Thus, Browett (1984: 169) claimed that although regions exist as material entities, "'the regional' is insignificant as a theoretical category" because, he professed, spatial form "cannot be regarded as an essential or central feature of society" (p. 170). In response, Smith (1986: 95–96) suggested that Browett had engaged in a "reverse fetishism" – that of completely ignoring the constitutive role of space within capitalist accumulation processes – and viewed the region like a Roman mosaic in which "[e]ach regional piece is made of very durable material, and once laid in place, it survives for a long time," such that although "the surface of each piece is continually rubbed and polished over the years by continual comings and goings, . . . the embedded pattern remains virtually unchanged." For Smith, Browett's region is essentially timeless and unchanging. By way of contrast, Smith (p. 96) argued that uneven geographical development is in fact central to how capitalism functions, an approach which gives the region much greater theoretical import – to carry the mosaic metaphor further, he contended that the dramatic economic restructuring taking place in the 1980s was

> akin to digging up the regional mosaic of past years and completely recasting a new pattern with new pieces made of different material and of different sizes . . . In short, it is not just a matter of breaking down regional boundaries but rather of transforming the very basis on which regions develop, their function, and the scale at which they exist.

Put another way, he sought to focus attention away from the nature and form of regions and to the process of regional formation, how this process is linked to the deep workings of capitalism, and how it, in turn, shapes the accumulation process. Such efforts by Smith and others to understand processes of regional formation were viewed as attempting to develop a "Reconstructed Regional Geography." Paradoxically, in some ways this approach had deep roots – in 1940, for instance, although avoiding the Marxist language, McCarty had argued that economic regions are important to theorize not because of their logical position within a scalar division of the world or because they are handy methodological units but because they are real entities created by market forces.

At the core of Smith's (1984) argument was the claim that the nature of regionalization had been transformed under capitalism. Thus, in the pre-modern era regionalization was largely determined by Nature – for instance, favorable climate and topography had shaped the geography of wool production in medieval England, which in turn had ensured the growth of market towns in East Anglia. However, as humans' capacity to "produce Nature" increased, economic regionalization was shaped less by climate and geology – as witnessed in the late nineteenth century by irrigation schemes turning California's Central Valley into a veritable cornucopia of agricultural production. Accordingly, Smith (1984: 115) argued, as capitalism developed, uneven development qua regionalization "based on accidents of uneven natural endowment [was] replaced by uneven development based more and more systematically upon the spatial differentiation of the social determinants of capital accumulation." Consequently, by the early post-Second World War era economic regionalization had come to be based "first and foremost" upon a socially determined territorial specialization reflective of the production process (p. 114). Moreover, whereas Kantian-inspired geographers considered the region "a geographic generalization" (James 1952: 199), for Smith it was a product of capitalist accumulation whose spatiality is defined by its functional coherence and the law of value, an approach which does not presuppose any generalization of conditions across space.

As empirical substantiation for such theoretical claims, Smith (1984: 115) examined how deindustrialization was reshaping the US's regional industrial geography and producing "a whole new configuration of regions." Specifically, he analyzed how the US's traditional northern industrial core had undergone a process of re-regionalization after 1945 (Smith and Dennis 1987). Based upon an examination of state-level manufacturing employment changes, he showed that in the immediate post-war period there was no statistically significant commonality of experience across the region. However, from the late 1940s to the mid-1970s there was a growing concurrence of experience as states began encountering similar employment change patterns. This reflected the emergence of a relatively internally coherent economic region out of the disparate experiences of the pre-war economic landscape – in other words, it highlighted the process of region creation and integration over this quarter century period. Conversely, from the mid-1970s until the mid-1980s this process seemed to work in reverse, as states' experiences began to diverge – whereas some went through significant employment loss, others actually added manufacturing jobs. This latter period was one in which the economic region that had coalesced in the previous two decades began to fragment, which suggests that the economic geography of the area was being re-regionalized.

Smith (1988) also looked at US political sectionalism, finding that periods in which local economic interests dominated Congressional voting patterns tended to be associated with economic depression whereas periods of economic expansion tended to exhibit greater regional commonality of interests. Equally, in the UK a regional political realignment reflective of underlying economic patterns had occurred as the booming South increasingly voted Conservative whilst other regions, especially those undergoing significant job loss, became more predictably Labour (Green 1988; Johnston et al. 1988). Based upon these findings, Smith argued for a materialist understanding of how regional scales of economic organization are made and unmade, suggesting that "[i]n the most abstract terms, we can conceive of regions as absolute economic spaces stabilized (however temporarily) in a wider sea of continually transforming relative space. They are geographical platforms of production, i.e., territorially discrete production systems; far from being static they are continually replacing and adding new planks as old ones become obsolete" (Smith and Dennis 1987: 168). Such a view, he argued (1988: 149), "should in no way be confused with the sterile efforts of traditional regional geography to 'define' the region sui generis," but was, instead, an effort to challenge idealist notions which see regions as mere conceptualizations and arbitrary spatial units.

Smith's approach presented a theoretically robust mechanism for understanding regionalization. However, for some it seemed too capital-centric and/or to downplay agency, either collective or individual. Accordingly, Herod (1997b) explored how US dockers had constructed a new regional economic geography of their industry through forcing new bargaining systems and wage and benefit structures upon their employers. For his part, Cooke (1985) examined how class practices in nineteenth-century South Wales had given the area a particular regional characteristic, suggesting that this created a center of strong working-class resistance to the dominant capitalist social relations of the time. Equally, whereas Reconstructed Regional Geography had initially been dominated by Marxist writers, who tended to focus on the economic aspects of regionalism, there soon emerged a call also to focus on "culture." Largely associated with the humanistic tradition in Geography (see Peet 1998), this approach concentrated on issues such as regional identification, with the region seen as "a specific set of cultural relations between a specific group and a particular place" (Paasi 1991: 240). From a non-Marxist perspective, Tuan (1982) contended that the landscape's regionalization and fragmentation are the result of individuals' sense of union with others. Others, such as Pred (1984), sought to advance a structurationist approach to bringing structure and agency together in an ongoing interaction, although as Gilbert (1988: 213) points out, whereas in Marxist approaches the region is understood as a product of specific social relations, in the interactionist/structurationist approaches it is seen more as the active context for such social relations' organization. Finally, it is significant to note that similar developments were occurring in francophone Geography, where geographers were exploring how the relationships of individuals and groups with differential access to power shaped processes of regional formation and how general and region-specific processes interact to create regional differentiation, although francophone geographers tended to focus more on the relationship between the region and group identity whereas anglophone geographers were more economically oriented (Gilbert 1988).

Such approaches to the region remained fundamentally topographical in nature – although they were conceptualized as the social products of economic and political processes and conflicts, regions were neverthe-less still largely viewed in areal terms. In the 1990s, however, as was the case with conceptualizations of the urban, much theorizing about the region begin to adopt a topological approach. Hence Thrift (1993: 94–95) opined that because of the "increasing mobility and fluidity" of contemporary capitalism in which place is "permanently in a state of enunciation, between addresses, always deferred," such that "space–time is no longer seen as bounded [but] is constantly compromised by the fact that what is outside can also be inside" (as when the local is viewed as the place where global forces are grounded whilst the global is seen as a collation of things local), it was becoming increasingly difficult to imagine cities and regions "as bounded space–times with definite surroundings, wheres and elsewheres." Likewise, Amin (2004: 33) argued that "globalisation and the general rise of a society of transnational flows and networks no longer allow a conceptualisation of place politics in terms of spatially bound processes and institutions." Consequently, instead of being conceived of as areal, he maintained that regions should be regarded as networked nodes, as "temporary placements of ever moving material and immanent geographies, as 'hauntings' of things that have moved on but left their mark ..., as situated moments in distanciated networks, as contoured products of the networks that cross a given place" (p. 34).

Some authors, however, critiqued the growing fascination with networks and the topological, suggesting that whilst a "topological concept of space may have the virtue of simplicity, and have some heuristic value, [it] is likely to obscure ... [efforts] to understand the difference that space makes" (Sayer 2004: 268). Others urged a "both/and," rather than "either/or," approach to the matter of whether regions should be considered areally or in rhizomatic, networked (i.e., topological) terms. Hence Morgan (2007: 1248) proffered that overcoming the

> debilitating binary division between territorial and relational geography [requires] recogniz[ing] that political space is bounded *and* porous: *bounded* because[, for instance,] politicians are held to account through the territorially defined ballot box, a prosaic but important reason why one should not be so dismissive of territorial politics; *porous* because people have multiple identities and they are becoming ever more mobile, spawning communities of relational connectivity that transcend territorial boundaries.

For MacLeod and Jones (2007: 1186), however, the upshot of such debates was "that the degree to which one interprets cities or regions as territorial and scalar or topological and networked really ought to remain an open question: a matter to be resolved *ex post* and empirically rather than *a priori* and theoretically."

CONCLUDING REMARKS

Clearly, the region has been a central concept within Geography, even as it has been conceived of in quite varied ways – at the beginning of the twentieth century primarily as a natural entity, by the mid-twentieth century increasingly in terms of culture, whilst both Spatial Science and Marxism saw regions chiefly as economic entities. Equally, as Gilbert (1988: 221) has observed, Regional Geography's focus has changed from "a regional analysis concerned mostly with the past" to being "a science of the present, a present understood in the light of the past," with such a shift "open[ing] up regional analysis to political action" as various social actors seek to craft regions in particular ways to suit their economic or political goals. However, there are some broader conceptual questions that come into play.

First, much of the narrative about Geography's intellectual history and where the region fits into this has relied for its coherence upon several powerful dualisms. Hence, the transitions in characterizations of the region during the twentieth century as generally represented rely for their coherence upon two sets of dualisms: a strict separation of Nature and Culture, and of Culture and Economy. As Smith (1984/1990) has shown, however, if we consider Nature a socially produced entity, then holding Nature and Culture apart as separate realms becomes impossible, an impossibility that has significant implications for how the transition from thinking of regions as defined by Nature to them as shaped by human activity ("Culture") is understood. By the same token, whereas the Economic and the Cultural have long been viewed as separate realms, recent scholarship (e.g., Thrift 2000) has argued that this distinction may not be sustainable. The representation of the 1950s as marking the replacement of a focus on the unique with a search for general laws also relies for its discursive power on the maintenance of a nomothetic–idiographic dualism. However, as Sayer (1989) has argued, this dualism is only sustainable if a Humean notion of causality is adopted, one wherein unique events cannot be theorized (because they are not part of a set of repeatable events) and laws can only be derived from examining events' regularities (Box 4.6). If, on the other hand, a realist approach is adopted, one in which the focus of analysis is upon determining the causal relations between natural and/or social objects rather than regu-larities in their spatial relationships (Sayer 1984), then the nomothetic–idiographic dualism is readily transcended. For his part, Stern (1992) maintained that the reductionist–holist dualism – is the region simply the sum of its parts or something greater? – could be avoided through adopting a "synergetics paradigm" which considers the region as a self-organizing complex system, an approach which also helps bridge the positivist/realist divide. Equally, Agnew (1999: 92–93) declared that the idealist–materialist dualism, at least as articulated with regard to regions, "rests on the total opposition that is drawn in contemporary geog-raphy between realism and constructionism as ontological positions" and that, in contrast to such a totalizing vision, regions actually "both reflect differences in the world and ideas about differences" and so cannot be reduced to the realm solely of either the material or the mental.

Box 4.6 POSITIVIST AND REALIST NOTIONS OF CAUSALITY

The Humean notion of causality (named for Scottish philosopher David Hume (1711–76)) is central to the positivist philosophy of science, upon which both traditional Regional Geography and Spatial Science drew. It argues that if Event B follows Action A frequently enough, Action A is said to cause Event B. Given this, it is impossible to theorize causality when dealing with unique events, for unique events, by definition, do not occur numerous times. This view of causality is quite different from the realist one, wherein there must be a causal mechanism linking Action A with Event B. In the latter perspective, then, it is not how often Event B follows Action A that is the determinant of causality but, rather, whether objects have *causal powers* and are necessarily connected in some way, such that one can influence another's behavior. Realist approaches also allow for contingent forces/factors to play important roles, such that the same causal mechanism may produce different results depending upon the context within which it operates – striking a match may cause an explosion in the presence of oxygen but will not in the presence of water – whilst different mechanisms may produce the same result (Sayer 1984). The realist view allows for unique events' causes to be theorized.

Destabilizing these dualisms, in other words, throws new light on Geography's disciplinary history.

Second, Entriken (1996) noted that much of the talk about a Reconstructed Regional Geography ignored physical geography. Although he himself did not make the distinction, it is useful here to think about such neglect in terms of both Physical Geography and physical geography. In the case of Anglo-American Physical Geography, this lack of attention may be related to the fact that a widespread approach since the 1960s in Geomorphology and, to an extent, in Biogeography, has been a focus upon small-scale surface processes and efforts to quantify the links between such processes and present-day landforms, in contrast to the approaches that dominated the first half of the century and which focused upon regional-scale narratives of landscape evolution (although recent developments in geochronological techniques and numerical modeling of landscape evolution at the regional scale may be serving to increase physical geographers' interest in regions) (Summerfield 2005). With regards to incorporating physical geography within debates over regions, if the notion of the social production of Nature is accepted, as Bourdieu (1991: 287) asserted, "the 'landscapes' or 'native soil' so dear to geographers are in fact inheritances, in other words, historical products of social determinants." Consequently, claims that physical geography was left out of efforts to develop a Reconstructed Regional Geography are perhaps less sustainable, as physical geography can be seen to be incorporated through recognizing that the "natural" landscape is itself largely a human creation. This is also the case if we adopt an actor-network theory approach, in which human and non-human objects (e.g., geological formations) are seen to interact.

Third, it is important to distinguish between methodological and ontological issues, for these can be confused when contemplating how to delineate regions. For instance, Hartshorne (1939: 296–97) argued that the term "region" (especially "natural region")

> implies that there is in . . . the real world . . . an unambiguous division of the earth surface and the problem is simply to recognize it correctly. No such division exists in reality [and] any attempt to divide the world involves subjective judgment, not in the determination of the limits of individual factors, but in deciding which of several factors is to be regarded as most important . . . The decision can only be subjective; the regions so constructed are in this sense arbitrarily imposed on reality.

For Kantians such as Hartshorne and Hart (1982), lines on maps illustrating regional boundaries are not assumed to bear any relation to underlying material realities. Ontologically, regions do not materially exist in the Kantian view and, consequently, there is little methodological problem in drawing regional boundaries – each boundary is simply where the individual geographer feels it should be and the only concern might be to minimize the degree of dissimilarity within the region. On the other hand, for those engaged in developing a Reconstructed Regional Geography, regions are real material entities and so accurately drawing regional boundaries raises significant methodological issues, given that these boundaries are assumed to bear some relation to entities existent in the world. Consequently, whereas both the Kantian and the Marxist approaches recognize that drawing regional boundaries on maps might bring methodological problems, they come at this issue from different perspectives based upon their ontological positions – for the Kantian, the issue is to develop boundaries which minimize internal differences, whereas for the Marxist the issue is to delineate accurately on a map boundaries that exist in reality.

A fourth issue is that of how the region has been made to perform – that is to say, how the idea of the region has been used to pursue particular agendas. Hence, drawing on the work of writers such as Butler (1990), Gibson (2001) has explored, in the context of industrial restructuring in Australia's Latrobe Valley, how the region was imagined as a place and how this imaginary was subsequently strategically deployed at various times within a rhetoric of the necessity of particular types of industrial restructuring. Likewise, MacLeod and Jones (2001) explored how "the North" in the UK has been constructed as an idea and how this has implications for national political discourse. Paasi (1991), meanwhile, conceived of the region as a historically contingent process always in the making and unmaking and suggested that discourses about particular regions can be deployed to encourage individuals to reproduce region-specific structures of expectations, whilst Linehan (2003: 118) showed how British regional development surveys in the twentieth century "helped . . . bring the region into a new space of governmentality for the operation of new powers of planning, regulation, statistical enumeration and representation," such that the region was made to play a particular role in understandings of economic development and underdevelopment that had "a profound effect on the shaping and imagining of the space economy." Lagendijk (2007), too, has suggested that the region can be viewed as a "performative concept" and that how the idea of the region is made to perform can have significant implications for how problems are perceived and thus how policies to address them are developed. It is important, however, to distinguish the performativity of the idea of the region from the Kantian view of the region as a mental image – in the former approach the focus is on the politics of the active construction of the idea of the region and how this affects social life, with no judgment made about the region's materiality, whereas in the latter the region is simply ideational.

Fifth, the recent engagement with post-structuralism has important implications for the practice of Regional Geography. In particular, early twentieth-century efforts at describing regions and their spatial extents did not have any explicit theory of observation and description – both activities were viewed as fairly unproblematic. However, more recently the practice of description and observation has been problematized. One of the earliest developments in this regard came from Paterson (1974: 5), who opined on the "logical impossibility of providing a complete regional description in verbal form." For her part, Pudup (1988) decried traditional Regional Geography's efforts to make (supposedly) theoretically neutral observations, a theme also taken up by Sayer (1984: 53), who explored how observation is "theory-laden" and how the "conceptually-saturated" character of observation can make it difficult to distinguish between what is observable and what is not. This latter fact has implications for how regions are conceived – must a region be visible in the landscape to be considered to exist and does the epistemological stance we bring to analyzing regions shape what we see them to be?

Finally, it is important to consider how the idea of the region might be used as a heuristic device without making any claims about the constitution of regions themselves. Thus, Haggett (1990: 78–83) outlined several ways in which regions as illuminatory devices might be used to help make sense of the world. First, they can be seen as exemplars to give "local substance to generalization, put flesh on the logical structure, provide a specific example to press home an argument." Second, they may be seen as anomalies or residuals, to illustrate how a local part of the Earth's surface may depart from more general statements about it, and in this regard they may play an important role in testing and reformulating general models. Third, regions may be studied as analogues, wherein the characteristics of one region are used to help understand another. Fourth, regions may be seen as modulators of broader trends/processes. Lastly, he suggests, regions can be seen as geographical "covering sets," spatial equivalents to the librarian's use of classification systems in which each area on the globe's surface can be fitted, with the rules of such regionalization being "comprehensive enough to ensure that the whole global region is covered, i.e. that there are no books which we cannot classify" (p. 83).

5

THE NATIONAL

[T]he nation-state is a power container whose administrative purview
corresponds exactly to its territorial delimitation.

(Giddens 1987: 172)

Generally, within Geographical (and other) scholarship, the spatial limits
of the national scale have been taken as coincident with the nation-state's
boundaries, to the point that relationships between nation-states are
typically viewed as being inter-national (i.e., between national entities).
The national scale, in other words, is seen to enclose the absolute spaces/
territorial expanses of a particular nation-state, even if few so-designated
national territorial units really fit the strict definition of nation-states
(political units consisting of an autonomous state that is populated
by a "nation" fairly homogenous in terms of language, ethnicity, and
history). Before considering how the national scale has been conceptualized, therefore, it is important to explore what we mean by a state.
In this regard, Tilly (1990: 1), noting that they have existed for about
8,000 years, has defined states as "coercion-wielding organizations that
are distinct from households and kinship groups and [which] exercise
clear priority in some respects over all other organizations within
substantial territories." It is significant, however, that at about the same
time that recognizable states were beginning to emerge so, too, were
cities. Indeed, for several millennia many of the states as defined above

were, to all intents and purposes, city-states – urban centers (usually focused upon some religious point of importance) which controlled a surrounding hinterland from which tribute was extracted (though see Hansen 1995 on problems of theorizing the city-state, at least in the Greek context). In fact, Tilly has argued (p. 2), "[t]hrough most of history, *national* states – states governing multiple contiguous regions and their cities by means of centralized, differentiated, and autonomous structures – have appeared only rarely . . . [and m]ost states have been non-national: empires, city-states, or something else." Only in the past few centuries, he averred (p. 3), "have national states mapped most of the world into their own mutually exclusive territories, including colonies." Consequently, although the nation-state's solidification as a political entity is typically seen to have begun with the 1648 signing of the Peace of Westphalia (notwithstanding the fact that entities with many of the characteristics of nation-states identified by Tilly existed prior to 1648, as in the case of the Meso-American Tarascan state (Pollard 1993)), it is really only since the end of World War II that "almost the entire world [has] come to be occupied by nominally independent states whose rulers recognize, more or less, each other's existence and right to exist." Nation-states as exclusive controllers of contiguous territory consisting of multiple cities and extending beyond the immediate hinterland of a capital city appear to be fairly recent inventions. Nevertheless, in terms of how the national scale has been conceptualized, it is this vision of the state as nation-state, rather than as city-state or as empire, that has dominated geographical scholarship.

Given this, in this chapter I address the rise of the national scale as an important scale of political organization, one often seen to have been formally secured by the Treaty of Westphalia and subsequently spread geographically through practices such as colonialism. The chapter also discusses debates concerning how the nation-state has been imagined as a discrete territorial unit and considers how issues of national sovereignty, and hence the power of the national scale, have been theorized within the context of contemporary "globalization" processes – many, for instance, have argued that globalization is encouraging the nation-state's "hollowing-out" and the undermining of the national scale of economic and political organization through the creation of supra-national entities such as the European Union.

ON THE NATIONAL SCALE'S ORIGINS

Perhaps the first issue to consider concerning the national scale is how nation-states as territories emerged historically. One thread of analysis has argued that they have their origins in military needs and the desire to secure a monopoly on the legal use of violence. This contention has a long history, stretching back at least to German sociologists Max Weber who, paraphrasing Russian revolutionary Leon Trotsky, maintained (1919/2004) that every state is founded on force, and Otto Hintze, who argued (1906/1975) that it is impossible to understand the state's development without taking into account how international rivalries, together with domestic conflicts between different interest groups and social classes, affect its internal structures. This is a contention also evident in the work of twentieth-century writers such as Foucault (1978/1997, 1978/2007), who argued that the state's principal role has been to facilitate the social disciplining of certain types of individuals (workers/ peasants, women, the insane) within particular territorial borders, even if how that has been done has changed historically. Hence, whereas during the medieval period in Europe the state was primarily concerned with controlling land and those who worked it and did so through customary and written law (e.g., serfdom), in the early Renaissance period its focus became the regulation of individuals through keeping various statistics (Catchpowle *et al.* 2004) (and here we should not forget that the word *statistics* first appeared in the late eighteenth century and derives from the Latin *status*, meaning "the state"). By the nineteenth century the state had increasingly come to focus upon controlling the population living within its territory rather than controlling its territory per se, which it did, in part, by creating particular modes of thinking – a *gouvernementalité*/governmentality (from *gouverner* [to govern] and *mentalité* [mode of thought]) – amongst the public. Of all writers, however, it is perhaps Tilly (1990) who has made the most sustained English-language theoretical argument in this regard (Badie and Birnbaum [1979] make a similar francophone argument) (Box 5.1).

At the heart of Tilly's case is the claim that new types of political units began to emerge in Europe after about the tenth century in response to states' need for resources to sustain their war-making activities. States could find these resources, he averred, in two kinds of settings. The first were what he called "capital-intensive" settings (typically, cities) in which resources were monetized and controlled by various economic

Box 5.1 BRIEF BIOGRAPHIES OF SEVERAL STATE THEORISTS

Otto Hintze (1861–1940) was born in Pyrzyce (Pyritz), which was then in Germany but is now in Poland. In 1880, he moved to Berlin to study for a doctorate under the historian Julius Weizsäcker, writing a dissertation on medieval institutions which would later be published as *Das Königtum Wilhelms von Holland* (*The Kingdom of William of Holland*). Ultimately he became Professor of Political, Constitutional, Administrative, and Economic History at the University of Berlin, though retired prematurely due to poor health. Under the supervision of Gustav von Schmoller, the leader of the "Younger" German Historical School of Economics which opposed axiomatic/deductive approaches to economics in favor of careful historically and geographically informed inductive study, Hintze was involved in editing the *Acta Borussica*, a collection of official documents detailing the history of the Electorate of Brandenburg and the Kingdom of Prussia that was published by the Prussian Archives. A historian of absolutism, Hintze argued that institutions related to the military and its finance were central in the creation of modern nation-states. In 1933, Hintze's wife Hedwig, a historian in her own right, lost her position as a lecturer and fled first to France and then to the Netherlands to escape Nazi persecution for her leftist beliefs and Jewish background. Although a more conservative Prussian, Hintze himself did not publish after the Nazis came to power and subsequently resigned from the Prussian Academy of Sciences and from his post as co-editor of the *Historische Zeitschrift* over his wife's treatment. For the last few years of his life, he was largely physically separated from his wife. In 1942, two years after his death, she committed suicide rather than be deported to a concentration camp. Several contemporary German historians believe Hintze to be the most significant historian of the German Empire and Weimar Republic.

Maximilian "Max" Weber (1864–1920) was a German sociologist and political theorist who was, amongst other things, the brother of the economist Alfred Weber, whose work on industrial location fundamentally shaped Economic Geography in the mid-twentieth century. Initially educated in Law at the University of Heidelberg, he later studied at the University of Berlin and at Göttingen. His law

continued

Box 5.1 continued

doctorate dissertation was on the legal history of medieval business organizations, whilst for his *Habilitation* thesis he wrote on Roman agrarian history, after which he joined the *Verein für Socialpolitik* ("Social Policy Association") and was put in charge of a significant study of how Eastern European farm laborers were migrating into places such as Prussia to take the jobs vacated by ethnic Germans who were leaving for the industrializing cities. Subsequently, he was appointed as a professor of Economics at the University of Freiberg and then at Heidelberg. His most famous work was arguably his book *The Protestant Ethic and the Spirit of Capitalism*, which argued that the rise of Protestantism – and especially Calvinism – in northern Europe facilitated the emergence of capitalism. Weber was involved in formal politics, trying to organize a party that would bring together social democrats and liberals in the years before the outbreak of World War I and serving on the Heidelberg worker and soldier council in 1918. He was also involved in working out the Treaty of Versailles and drafting the Weimar Republic's Constitution. After World War I, he taught in Vienna and then Munich, where he was head of the first Institute of Sociology at any German university. He caught the Spanish Flu and died of pneumonia during the epidemic.

Paul-Michel "Michel" Foucault (1926–84) was born to a prominent family in Poitiers, France. He was educated at the Jesuit Collège Saint-Stanislas during World War II, when Poitiers was controlled by the forces of Vichy France, and then at the École Normale Supérieure in Paris, where he was influenced by Marxist and existentialist ideas. He earned degrees in psychology and philosophy and joined the French Communist Party in 1950, though left after only three years. His mentor at this time was Louis Althusser. He held academic positions at the Université Lille Nord de France, the University of Uppsala (Sweden), Warsaw University, and the University of Hamburg. In 1960 he took up a post in philosophy at the University of Clermont-Ferrand, completing his doctorate in 1961 with essays on the history of madness in Europe and on Kant. At Clermont-Ferrand he met Daniel Defert, a philosopher who would be his lover for some two decades. He taught briefly at the University of Tunis and was

continued

Box 5.1 continued

then appointed by the French government to head the Philosophy Department at the new University of Paris VIII, which had been founded in the wake of the 1968 student riots. Engaging in sit-ins of university buildings with students, Foucault soon thereafter accepted a post at the prestigious Collège de France. He also taught periodically in the United States. Foucault's work focused upon issues of power and social control of the body, and he wrote extensively on matters of the history of sexuality, of punishment, and of knowledge.

Charles Tilly (1929–2008) was born near Chicago and educated at Harvard, Oxford, and the Catholic University of Angers, France. After serving in the US Navy's amphibious forces during the Korean War and receiving a doctorate in Sociology, he held teaching positions at the University of Delaware, Harvard, Michigan, Toronto, the New School in New York City, and Columbia University, where he finished his career as the Joseph L. Buttenwieser Professor of Social Science. He was a member of the National Academy of Sciences and of several other prestigious similar organizations, a Guggenheim Fellow, and a fellow of the German Marshall Fund. In 2008 he received the Albert O. Hirschman Award from the Social Science Research Council. He was awarded the *Chevalier de l'Ordre des Palmes Académiques* (Knight of the Order of Academic Palms) by the French government for his work, which was principally on state formation, social movements, and methodological issues. Tilly was a prolific writer, publishing more than 600 articles and 51 books and monographs.

actors who were often involved in some kind of value-added manufacturing or materials processing. The second were in what he called "coercive-intensive" settings (typically, rural areas) in which landlords who controlled territory rich in grain, timber or other raw materials relied upon coercion and intimidation to control those resources and extract wealth through their production, refining, and sale. As a result of both the tensions and the overlap between these two groups, together with the fact that these tendencies toward capital accumulation and coercion were distributed in a geographically uneven fashion across the economic and political landscape, Tilly suggested that three main types of political units emerged: i) coercive-rich/capital-poor tribute-taking empires such as Russia and that of the Ottomans, in which "rulers squeezed the means of war from their own populations and others they conquered, building massive structures of extraction in the process" (Tilly 1990: 30); ii) systems of fragmented sovereignty, such as city-states and urban federations that were capital-intensive but coercive-poor and in which rulers relied upon compacts with capitalists and merchants to rent or purchase military force and "thereby warred without building vast permanent state structures" (p. 30) (places such as Genoa, Dubrovnik, and the city-states of the Dutch Republic would be emblematic of such entities); and iii) states such as Britain and France, which were somewhere between these two extremes, wherein "[h]olders of capital and coercion interacted on terms of relative equality [and] produced full-fledged national states earlier than the coercion-intensive and capital-intensive modes did" (p. 30).

Certainly, Tilly did not suggest that war has been the only factor involved in nation-state creation – others have also been important historically. Moreover, such an argument is not to say that state formation for such purposes was necessarily a strategy thought out a priori, as "[r]arely did Europe's princes have in mind a precise model of the sort of state they were producing, and even more rarely did they act efficiently to produce such a model state" (Tilly 1990: 25). Additionally, states' principal institutions (treasuries, courts, central administrations) often developed on an ad hoc basis in response to various crises (often, military threats from outside or various internal insurrections and rebellions), and other states' actions frequently affected the path of change. Equally, although he focused upon European states' genesis, Tilly pointed out several major differences between how European states emerged historically and how the newly decolonizing states of Africa and Latin America did. Nevertheless, he suggested, taken as a whole, a principal reason for nation-states' emergence in Europe in the early Middle Ages was the increasing costs of warfare for control of territory and resources brought about by the growing use of gunpowder and professional armies, which in turn necessitated the growth of organizations capable of more effectively extracting taxes from the citizenry (see also Tilly 1985). Although Tilly's ideas have been critiqued by some (e.g., Goldstone 1991, Hobden 2001, Tarrow 2004), they have also been adapted to more modern processes of nation-state formation. Hence, in her analysis of the Nagorno-Karabakh conflict and state-making in Armenia in the early 1990s, Papazian (2008) points out that whilst Tilly's formulation is derived from historically and geographically quite different sets of events, his argument is still helpful in understanding more contemporary developments where war has often been a companion to state formation, including the emergence of states out of the hulk of the old Soviet Union or other empires (see also Blom, no date).

Tilly, of course, is not the first writer to contemplate the nation-state. Thus, the classic conservative position, articulated by Thomas Hobbes (1651/2002), interpreted the nation-state as having emerged to counter what Hobbes saw as humans' inherent ruthlessness and desire to pursue self-interest without regard for others. Without some restraining entity, Hobbes suggested, such innate nature would lead to permanent conflict between people. What was needed was a form of sovereign power – either a single person (his preference) or an entity such as a parliament – which could legitimately enforce order within society so that individuals could do what they wanted without interference by others. For Hobbes, the state emerged out of an agreement made between the sovereign and the people – the latter agreed to cede rights to the former and to give up the unfettered pursuit of their individual rights so as to avoid social chaos in return for protection and order. The result was a commonwealth in which individuals tempered their baser instincts through creating networks of associations amongst themselves but within the boundaries laid down by the sovereign. Paradoxically, perhaps, the classic liberal perspective on the state – as articulated by people such as John Locke – likewise suggested that there should be a social contract between the people and their rulers. The difference, however, lay in Locke's view of human nature. Specifically, Locke saw humans as essentially good and that the state was needed to allow people to fulfill their potential by removing barriers to their so doing. Hence, he argued (1690/1967) that the state developed out of people's desire for a neutral judge who could protect their lives, liberty, and possessions, rather than the need for a sovereign to stop them from fighting each other. For his part, Rousseau (1762/1935) maintained that the state materialized to ensure individual rights that had been secured over time as part of the social contract and that it could only really emerge once the consent of the governed had been secured.

Marx and Engels (1848/1948: 11), on the other hand, viewed the nation-state quite differently, seeing it as the product of a more general class struggle and an instrument through which one class dominates another – they famously referred to the state under capitalism as "a committee for managing the common affairs of the whole bourgeoisie." However, despite seeing its emergence as a necessary correlate to one class's securing of the surplus labor of another, they were also somewhat ambivalent with regard to it, deeming it as capable of being controlled by either workers or capitalists, depending upon the status of the class struggle between them (for more on Marxist-inspired theories of the state, see Jessop 1982; also, Lenin 1902/1988, 1917/1932). Drawing upon Marxist theory, Smith (1984/1990) conceptualized the nation-state, and thus the national scale, as growing out of tensions between capital's need, on the one hand, to equalize conditions across the Earth's surface through universalizing the law of value and, on the other, to differentiate the landscape into specific absolute spaces for accumulation purposes. Thus, he averred (pp. 142–2), the "impetus for the production of this scale [came] from the circulation of capital, more specifically from the dictates of competition between different capitals in the world market," which resulted in a "hierarchy of nationally based laws of value more or less integrated within a larger international law of value." Although he recognized that various other types of state clearly predated capitalism's birth, Smith stressed that the economic forces of capitalist accumulation dramatically reworked their nature. Hence, whereas capital-ism inherited "a geographical structure of city-states, duchies, kingdoms and the like – localized absolute spaces under the control of pre-capitalist states – . . . it transform[ed] what it inherit[ed and] generally combine[d] a number of these smaller states into . . . nation-state[s]." If the national scale is created out of the needs for capital accumulation, however, for Smith so, too, are the sizes of the absolute spaces that national boundaries encircle. Hence, the "geographical extent of the nation-state is constrained on the low end by the need to control a sufficiently large market (for labour and commodities) to fuel accumulation. At the high end of the scale, a nation-state that is too large finds it difficult to maintain political control over its entire territory." Although the "actual determination of the limits to this scale . . . is politically determined by a series of historical deals, compromises and wars," it is nevertheless "provoked," he con-tended, by the dialectic of equalization and differentiation. The result is, as Brewer (1990: 266) put it, that "[c]apitalism and the nation state [have grown] up together."

Whilst Marx and Engels believed that the state would have to be captured – either peacefully or violently – by workers and turned to their advantage as part of a socialist revolution until its repressive elements gradually withered away and it was left as, essentially, a worker-controlled bureaucracy for organizing social life, many anarchists have seen the state as having emerged to protect the private ownership of the means of production which is at the heart of the capitalist system and so, they have argued, there is no choice but to destroy it if capitalism is to be overcome (Bakunin 1867–72/1950; Meltzer 1996). On the other hand, anarchists of a more libertarian bent have suggested that "[c]apitalism is the fullest expression of anarchism, and anarchism is the fullest expression of capitalism," such that "[t]rue anarchism will be capitalism, and true capitalism will be anarchism" (Rothbard 1972). Significantly, these commentators have argued that the state is essentially a *kleptocracy* that developed out of the desire of various ruling classes to control land-use and the labor of those who worked it and that the state greatly limits the competition that naturally emerges out of capitalism and should, therefore, be overthrown.

More recently, feminist scholars have also theorized the state in various ways. So-called liberal feminists have viewed it as essentially a neutral adjudicator between various interest groups. Although the state has historically been dominated and controlled by men, they argue, there is nothing to stop women from securing command over it, such that having more women in elected positions will inevitably lead to policies that push women's interests. The state's neutrality is illustrated most clearly when women and men are treated by it as equal citizens. Radical feminists, on the other hand, have tended to stress the state's inherently patriarchal nature, arguing that all states – even those that profess to treat men and women equally – are de facto patriarchal because patriarchy is a global phenomenon and, at the end of the day, states that are created out of patriarchal conditions will serve to reinforce such patriarchy. Socialist feminists have offered yet another perspective, seeing the state as foremost a product of capitalist social relations in which women's subjugation plays a central role in maintaining capitalism. Hence, they see patriarchal relations within the state – as manifested, say, through the state's role in reproducing ideologies and material practices wherein women's "proper" place is seen as being in the home raising children – as primarily about sustaining capitalist accumulation and the reproduction of capitalist social relations (for critiques of aspects of these positions, see Kantola 2006). For their part, Marston and Smith (2001) have suggested that US women's movements of the nineteenth century played key roles in state formation by engaging in movements to secure social welfare and various municipal reform programs – "the modern state was remade," they argue, "with new responsibilities and transformed roles in direct response to the prolonged political activism of urban women, organized around a discourse of domesticity and maternalism" (p. 617). Finally, some queer theorists have argued that the state emerged as an entity to ensure that property inheritance occurred along patriarchal lines (Nast 2005) and that the patriarchal state both reproduces notions of heteronormativity – through, for example, "family" social policy that naturalizes heterosexuality and presents homosexuality as abnormal – but also reproduces itself through dividing the absolute space of the Earth into territorial containers through which various heteronormative binaries of identity (foreign/non-foreign; domestic/non-domestic) can be constructed and deployed (Mosse 1988; Morgan 2000; Puar 2007; Scott 2007).

Despite these various theoretical approaches' differences, they have all generally seen the state in national territorial terms and circumscribed spatially by borders which are viewed as representing the geographical extent of the national scale. Implicitly, they tend to assume that the 1648 Peace of Westphalia marked a significant event horizon in the state's development into its modern, nation-state form through confirmation of its power over the territory contained within it and, thus, greater respect for ideas concerning states' sovereignty and territorial integrity (Box 5.2). Consequently, it has generally been argued that the Peace marks the beginning of a growing coincidence between nations and nation-states. Indeed, the power of this representation is revealed in the widely made claims – particularly from neoliberal commentators (e.g., Ohmae 1995), but from others too – that the nation-state is being undermined by the forces of globalization and that a post-Westphalian system of territorial organization is emerging, one in which nation-states' roles as containers of economic and political life are being challenged (Herod 2009). Nevertheless, despite the vigor with which such arguments are often made, in fact the relationship between nations and the nation-state has been much more complex historically and geographically. Hence Hobsbawm (1990) has argued that the French nation-state was founded prior to the population of the national territory developing a sense of being "the French people" and that such a people was only subsequently created through repression of regional dialects and languages and their supplanting with a standardized "national" French

Box 5.2 THE PEACE OF WESTPHALIA

The Peace of Westphalia refers to two treaties signed in 1648, those of Münster and Osnabrück. The Treaty of Münster ended a war between Spain and its Dutch subjects and resulted in the eventual creation of an independent Dutch Republic. The Treaty of Osnabrück ended a series of religious and other conflicts within the Holy Roman Empire. The Peace established a number of tenets which would serve as the basis for modern nation-states, including the concept of a sovereign state governed by an authority having control over a particular territory and population and not subject to control by another, the right to self-determination (manifested in the acceptance that each ruler could determine the religion of his/her own state according to the principle of *cuius regio, eius religio* ["Whose realm, his religion"]), and the recognition that such states' territorial integrity should be respected by others, even in military defeat. This was a quite different view of territory than had existed until then, when, for instance, various cities or provinces might be given by one ruler to another, perhaps as a wedding present. The Peace of Westphalia also removed several barriers to trade.

language. In Germany, however, it was the self-awareness that something called "the German people" existed that drove the German nation-state's ensuing formation. Equally, Teschke (2003) has argued that whilst 1648 is often interpreted as marking the historical origin point of modern international relations, in fact European international relations remained mired in dynastic and absolutist politics, themselves rooted in feudal property regimes, until the beginning of the industrial revolution in Britain, after which the British model was exported to Continental Europe and beyond, often through colonialism (Osiander (2001) makes a similar argument). This process of nation-state development, Teschke indicates, was both historically and geographically uneven, and was not really completed in Europe until World War I, two and a half centuries after the Peace of Westphalia was agreed.

In similar fashion, Wiebe (1967) suggests that in the United States the nation-state really began to crystallize in the late nineteenth/early twentieth century as the frontier's closing allowed people to "savo[r] the word 'nation' in th[e] sense of a continent conquered and tamed" (p. 11) and as the massive social transformations that were taking place (mass immigration, urbanization, industrialization) undermined the long-standing values of small-town life and increasingly allowed the emergence of a sense of national order as the country was increasingly bound together by economic, demographic, and other processes and as new, usually nationally organized, institutions emerged to manage these processes. For his part, Anderson (1983) has argued that in the West nation-states, particularly as "territorial expressions of national capitalism" (the phrase is Mac Laughlin's (2001: 100)), really only emerged in the context of a developing modernity and that, consequently, the eighteenth and nineteenth centuries are their more appropriate historical starting points.

These disputations raise important questions about how nation-states' origins have been understood – especially whether 1648 is seen as the key date or whether some latter sets of dates, say 1815 and the Congress of Vienna, are more appropriate – and how nation-states have come to be discursively constructed as powerful geographical containers of both peoples and economies within "a world of absolute spaces and explicit, non-overlapping boundaries" (Warf 2009: 64) that is quite different from the feudal states of the Middle Ages with their often fairly evanescent boundaries which could change as land was added or lost through, for instance, a royal marriage. Although much is dependent in these claims upon the break-up in Europe of multi-ethnic empires such as the Holy Roman Empire or the Austro-Hungarian Empire and the amalgamation of myriad city-states, principalities, and duchies into nation-states, part of the story also lies in developments in the visual representation of nation-states, especially cartographically, for map-making has been a powerful tool in reinforcing the sense of a national scale and control over territory. Hence, although mapping had always been an important means of illus-trating the state's power over territory (the Romans, for instance, developed complex cadastral systems for tax collection purposes), it took on par-ticular importance in early modern Europe, where not only was mapping a way to delineate accurately national territory to establish/reinforce sovereignty, but it was also tied up with the ideas of political reform and rational thinking emerging during the Age of Reason – accurate mapping was seen as central to providing for better economic planning and political representation, for instance (Buisseret 1992). Mapping was also intimately embroiled in practices of imperialism, as European powers took representational control of conquered territories and shaped their cartographic depiction to their own goals. The English Crown, for example, used mapping in its conquest of Ireland to turn a landscape that was culturally illegible and a "disorientating space" to English administrators into a "legible, ordered imperial territory," in the process "invent[ing] 'Ireland' as a geographical and discursive entity" (Ó Tuathail 1996: 4) – by encouraging the disparate clan members increasingly to see themselves as part of a unified "imagined community" (Anderson 1983), an Irish nation, English colonialism, in other words, helped bring about the Irish nation-state as a territorial entity encompassing a group of people with a sense of common heritage. As Wolfart (2008: 4) suggests, in cases of colonial mapping invariably what had begun as a cartographic "exercise in demarcating a border and asserting administrative territorial control . . . led to new conceptualisations of sovereignty and territoriality. Space came to be integrally tied to territorial sovereignty and regional identity, and mapping became a key instrument in the construction of territorial nation-states," a process in which indigenous populations were often quite literally mapped out of the picture (Bassett 1994; Brealey 1995; Godlewska 1995; Ryan 1996). Significantly, these developments in representation and their effects on how the nation-state and the national scale have been conceptualized were, according to McLuhan (1969/1997: 233; also Anderson 1983), an extension of earlier connections between visual media and nationalism going back to the beginnings of the printing press, when greater familiarity with a standard typography "enabled every literate man to *see* his mother tongue analytically as a uniform entity." The printing press "by spreading mass-produced books and printed matter across Europe, turned the vernacular regional languages of the day into uniform closed systems of national languages . . . and gave birth to the entire concept of nationalism."

IMAGINING THE NATIONAL IN GEOGRAPHIC DISCOURSE

Having explored how nation-states' origins have been understood, together with some of the debates concerning when they emerged historically as relatively geographically coherent entities embodying a national scale of social organization, in this section I outline three common ways in which the nation-state as a scaled territorial entity has been represented in the geographic literature: in terms of an organismic analogy; as a spatial container of various aspects of social life; and, more recently, as a networked entity.

The organismic analogy

The idea that the nation-state can be thought of as a living entity – as an organism – has a long history. Thus, in his *Republic* Plato (2000: 105) suggested that the human faculties of reason, courage, and desire could be best understood through their representation both as parts of the body (the head, the heart, and the stomach or genitals) and as various social groupings which make up the state (its ruling and intellectual class, its warrior class, and its merchant class). Similar arguments were made by other writers from the ancient world, including Aristotle, Cicero, Livy, Seneca, and St. Paul (Coker 1910: 13), whilst in the Middle Ages John of Salisbury (1159/1990) viewed various parts of the state as like body parts. Five centuries later, Hobbes (1651/2002: 9–10), in *Leviathan* (in which the frontispiece showed a picture of the sovereign literally embodying the state), maintained that

> For by art is created that great LEVIATHAN called a COMMON-WEALTH, or STATE . . . which is but an artificial man, though of greater stature and strength than the natural, for whose protection and defense it was intended; and in which the sovereignty is an artificial *soul*, as giving life and motion to the whole body. The *magistrates* and other *officers* of judicature and execution [are] artificial *joints*. *Reward* and *punishment* (by which fastened to the seat of the sovereignty, every joint and member is moved to perform his duty) are the *nerves* that do the same in the body natural. The *wealth* and *riches* of all the particular members are the *strength*. *Salus populi* (the *people's safety*) [is] its *business*. *Counsellors*, by whom all things needful for it to know are suggested unto it, are the *memory*. *Equity* and laws [are] an artificial *reason* and *will*. *Concord* [is] *health*. *Sedition* [is] *sickness*. And *civil war* [is] *death*. Lastly, the *pacts* and *covenants* by which the parts of this body politic were at first made, set together, and united, resemble that *fiat*, or the *let us make man*, pronounced by God in the Creation.

In similar fashion, the French revolutionaries of the late eighteenth century often saw the Revolution as a process whereby local loyalties to regions such as Picardy or Brittany would be replaced by allegiance to a new, national political body – France. Within such discourses biological and corporeal metaphors were frequently used to describe the citizen's relationship with the new entity, such that patriotic devotion was seen to be, as one revolutionary pamphlet put it, for "the body politic what the circulation of the blood is for the veins of the human body" (Birnbaum 2001: 45) (Box 5.3).

Although such ideas have surfaced numerous times throughout history, one of the earliest modern writers to think much about the state systematically in these terms was Hegel, who viewed it as embodying the spirit (*geist*) of a people and their society. For him, the progression

---

Box 5.3 CONSTRUCTING A NATIONAL SCALE OF IDENTITY IN POST-REVOLUTIONARY FRANCE

One of the key goals of the revolutionaries who seized power in 1789 was to replace what they saw as parochial localist sympathies through creating a new national sense of Frenchness unconnected to France's main institutions until that point – the Monarchy and the Church. Such efforts were manifest in several ways – executing the king and establishing a republic, abolishing the aristocracy, establishing a new metric system of weights and measures, and devising a new calendar unconnected to the religious festivals and dating system of the Catholic Church, amongst other things. As a way to erase old regional loyalties and create a sense of nationalism, the revolutionaries pursued two goals in particular. First, they divided the country up into new, "rational" political units of roughly equal size, thereby breaking up the long-standing regionalization of *ancien régime* France which had shaped people's sense of local identity. Second, they pursued a policy of suppressing regional languages such as Provençal and Gascon and encouraging the use of French as a national language, even though it was spoken well by only about 13 percent of the population and not at all by half of it (Hobsbawm 1990). Such efforts continued into the nineteenth century as the French educational system became increasingly centralized. The erasure of regional languages was seen as central in fostering a national identity amongst the population.

of history through the process of thesis–antithesis–synthesis is a process in which ideas drive social and political developments – a distinct
contrast to Marx's argument that struggles over material conditions do
so – as societies seek to achieve a form which satisfies their most
fundamental desires (Box 5.4). In such a formulation, the *geist* takes on
spatial form, being contained within the geographical boundaries of the
nation-state. However, although the state is an embodiment of the spirit
of the age, different states can be humanity's developmental leaders
at different historical moments – once some nations have reached a
particular cultural or political plateau, they might sit back to contemplate
their achievements and thus allow other states to take what they had
learned and use that as a launching pad for their own development. For
Hegel, the state is the highest form of social being and other social units
– community, the family, etc. – only have meaning in relation to it. In
this formulation, the state has a purpose and will of its own that is
independent of the individuals living within it – it is "superorganic," an
entity greater than the sum of the individuals making it up and one which
behaves according to its own internal laws.

If Hegel viewed the state in superorganic terms as a manifestation of
God's efforts to understand himself, with the Absolute Spirit of the age
taken as God and the state as "the Divine Idea as it exists on Earth" (Hegel
1837/2004: 39), then a number of late nineteenth-century writers
envisioned the nation-state in more explicitly organismic terms as a kind
of amoeba which could expand into other territories and absorb the
weaker states found there. Such views took earlier understandings of the
state as an organic, if somewhat metaphysical, embodiment of various
values one step further to see it in terms of an actual organism, with
some viewing the state in psychic terms (considering it to possess various
fundamental attributes of human personality and/or to have parallels
to the ages of a human – being childlike, mature, old, etc.) and others
viewing it in biological terms (with various state organs playing essentially the same role as the equivalent organs in animals and plants) (Coker
1910). Such ideas were explored in several social sciences and countries – French geographer Vidal de la Blache (1918/1926: 112), for
instance, argued that "the European organism is in such a condition that
its motor nerves today are very active, even to the farthest extremities of
its various members." However, the idea of the state as an organism has
arguably been greatest in German writing (Coker 1910 provides a good

Box 5.4 DIALECTICS

Georg Wilhelm Friedrich Hegel (1770–1831) was a German idealist philosopher who drew on the works of, amongst others, Immanuel Kant to argue that how we understand objects in the world is the result of how we perceive them to be, rather than the result of any inherent properties they possess. Consequently, for him any order discovered in the world comes from the categorizations placed on it by the mind, rather than the playing out of attributes which such objects have due to their own material existence. Within this context, Hegel suggested that what drives History is ideas – the belief that ideas of freedom will eventually win out over those of tyranny, for instance. Hegel argued that History unfolds in a dialectical manner as new ideas are developed and replace outdated ones. Hence, the famous dialectic of thesis–antithesis–synthesis named for Hegel may be summarized as the situation in which an extant historical condition (thesis), such as a repressive government, produces its own antithesis (say, a social movement filled with ideas about liberty), which overthrows the repressive government but often reconciles the revolutionary present with the repressive past (synthesis) by adopting some of the overthrown regime's attributes (revolutionary governments, for example, often become repressive as they seek to limit opposition from the former government's supporters). Although Hegel's work has been interpreted by some (particularly the French philosopher Alexandre Kojève (1947: 464–67)) to argue for the possibility of an "End of History" when a society reaches humanity's "ultimate objective" (freedom guaranteed within the liberal democratic state), Dale (2006) and others have maintained that Hegel adopted no such argument. Significantly, though, the idea of an End of History as a Hegelian concept has been popularized by neoconservative Francis Fukuyama (1992) in his description of the Soviet Union's collapse and the "victory" of Western liberal democratic states and market economies.

Karl Marx took Hegel's ideas about the dialectic but famously "turned Hegel on his head." Consequently, whereas Hegel had argued that ideas drive History forward, Marx adopted a materialist view to argue that it is, instead, the material conditions of existence that do so and that the dominant ideas in any society merely reflect who has

continued

Box 5.4 continued

the economic and social power to shape its members' worldview. For Marx, the dialectic is driven not by conflicts over ideas whose relative merits are weighed in some sort of transcendental evaluation but by the working out of contradictions inherent in a society's extant conditions – in the case of capitalism, the fact that the producers of wealth (the proletariat) do not own the product of their labor.

---

overview), with perhaps the most influential writer in this regard being the German geographer Friedrich Ratzel who, significantly, had trained as a zoologist.

For Ratzel, who was quite enamored of Darwinian metaphors and images and sought to apply Darwin's work to explain the social world, the main weakness in the latter's analysis was its failure to take account of the spatial dimension of competition between different forms of life (Heffernan 2000). Consequently, in his 1897 book *Political Geography*, Ratzel outlined his theory of how the nation-state behaves like an organism and needs to expand geographically if it is to secure its future. Arguing that all organisms are involved in a struggle for *lebensraum* (living space), he maintained that this shapes their behavior. Whilst he was quite careful to suggest that his analogy not be taken literally, Smith (1980: 54) avers that he did tend to employ it "as though it were a law of nature" (though Dikshit (1999: 22) suggests Ratzel's use of the word "law" refers more to tendencies than to absolutes). For Ratzel, the relationship between a people and the land was what gave a nation-state its internal coherence and the two together formed a kind of living organism which sought to fill out its "natural region" and to overcome artificial boundaries imposed to protect weaker states. Hence, states were territorial both in form and in character, seeking both to occupy their own natural spaces and to expand into the spaces of other, weaker states (see Bassin 1987). Given his use of Darwinian imagery, Ó Tuathail (1996: 28) has argued that Ratzel's view marked a "biologization" of global space.

Following Ratzel's death in 1904, it was arguably the Swedish geographer Rudolf Kjellén who most further developed the idea of the state as an organism, drawing upon Ratzel's ideas but also expanding upon them. Whereas Ratzel had focused upon the nation-state's spatial configuration and thus his was a geographically determinist position, Kjellén (who coined the term *Geopolitik* to describe Ratzel's position that the physical environment influenced how states behaved) also included an analysis of population (*Demopolitik*), economy (*Wirtschaftspolitik*), society (*Soziopolitik*), and power (*Herrschaftspolitik*) (Herb 1997). If Ratzel was somewhat ambivalent about the state's biological nature, however, Kjellén saw the state as quite literally an organism – as Tunander (2001: 453) puts it, Kjellén's "organic view of the state was an attempt to regard the state as an independent object of study with its own dynamic and logic, power and will, an organic unity of land and people, an organism with body and soul, a personality on the international stage." In turn, German geographer Karl Haushofer was greatly influenced by Kjellén's ideas of the state as a biological organism and by Ratzel's notion of *Lebensraum* (Crone 1951: 44). In particular, although there were problems for Haushofer with adopting the notion of *Lebensraum* – for instance, whereas Ratzel had envisioned *Lebensraum* as a requirement for agrarian societies, Haushofer sought to apply the idea to an industrialized Germany to justify the latter's geographical ambitions in the 1930s (Smith 1980) – the idea that nation-states are dynamic organisms with natural boundaries to which they need to expand geographically, rather than entities with fixed, juridically determined boundaries which might be too confining, appealed to him (Tunander 2001). In particular, the idea of an organismic state provided legitimacy to German desires to escape the constraints of the Versailles Treaty, whose artificially imposed boundaries were seen to be constraining the German state's "natural urges," and provided rhetorical fuel for the Nazis. Indeed, the idea of the nation-state as an organism driven by a biological will to survive has been quite common in fascist ideology – the Mussolini government's Charter of Labor, for instance, declared the Italian nation to be "an organism having ends, life, and means of action superior to those of the separate individuals or groups of individuals which compose it" (quoted in Randall 1976: 663).

Given how they were associated with Nazi ideology, after 1945 ideas about the state being organismic largely died in Europe, or at least moved into the shadows. Nevertheless, they did carry on in Latin America, particularly in countries ruled by military governments. For instance, in Brazil General Golbery do Couto e Silva, an anti-communist whose book *Geopolítica do Brasil* (*Brazilian Geopolitics*) had an important impact on geopolitical thinking in the post-war era, adopted Ratzel's and Kjellén's ideas concerning the nation-state as a spatial organism. However, he extended their ideas to suggest that the nation-state's body politic might not only face threats from outside but might also have to deal with "cancerous cells" internally, such as communist groups or regional independence movements (Hepple 2004). Similar ideas spread to other South American dictatorships, including Argentina and Chile, with the growing use of metaphors regarding cancerous growths within the nation-state resulting from the fact that knowledge of cancer and how it spreads within bodies was much greater during the post-1945 era when people such as Golbery were writing than it was when Ratzel and Kjellén were initially writing (Hepple 1992).

Despite the disrepute into which organismic analogies fell in Europe as a result of them having been widely seen as ideological justifications for fascist states' expansionary policies, in more recent years the metaphor of the state as a biological organism appears to have enjoyed something of a revival. For example, Thornborrow (1993: 117) has shown how Europe as a geographical region has increasingly come to be represented in anthropomorphological terms, particularly in France. Likewise, Luoma-aho (2002a, 2002b) has illustrated how the conflict in the Balkans in the 1990s was represented within European security discourse in bio-medical terms as a threat to Europe, which was itself imagined as a territorial organism. The cause of such danger, Luoma-aho (p. 119) argued, was a malignant nationalism which was portrayed as in need of immediate excising, cancer-like, for "unless treated, [it] could spread to other states in eastern Europe, and even involve the member states of the EU." Balkan nationalism, in other words, threatened to kill the nation-states of Europe. Finally, it is important to recognize that although the nation-state has often been described in organismic terms, the specifics of particular organismic metaphors change over time as understandings of the body change or as the referents upon which the metaphor draws gain or lose currency. Thus, the embryologist and comparative anatomist Oscar Hertwig (1922: 54–55, quoted in Gerschenkron 1974: 435) argued in the early twentieth century that

> telegraph and telephone wires can be . . . described as the nerves of the state, because in the same fashion as the nerves in an organism they directly connect the individual independent members and carry messages over long distances. The telegraph and telephone offices are the ganglions of the state. In the same way, roads, railroads, and

> canals with the relevant plant establishments play a similar role as
> the lymphatic vessels in a corporeal organism. The individual in the
> organism is the cell; in the state it is the man.

However, once the wireless began to replace the telegraph, the metaphor
of telegraph wires representing the state's nerves came to have less hold
over the imagination.

The nation-state as container

If the national scale has often been imagined in organismic terms, with
various national boundaries serving as "the skin of the state" (Hepple
2004: 360), then a second – and closely related – way in which it has
been envisaged has been as a container for various social attributes. Parkin
(1979: 121) has argued that German writers of the late nineteenth/early
twentieth centuries in particular viewed the state in terms of a container
which could "defend the boundaries of the cultural community against
erosion or assault by powerful neighbors," suggesting that for authors
such as Weber this may have reflected preoccupation with the country's
recent unification and protecting its national borders (especially to the
East, where the German/non-German geographical division seemed less
clearly defined than to the West, with its sharp break between Germanic
and Romance language and culture). Although organicist conceptions
often viewed the state as a cultural container (e.g., as the German nation's
container, such that its boundaries needed to grow until all German
populations were held within it), in such a conception the nation-state
was also viewed as a container for economic practices – it was seen as
"defining the basic unit for economic transactions" and there was
assumed to be an "organic unity" to the national economy (Agnew and
Corbridge 1995: 61). Indeed, it would be almost impossible to imagine
nation-states as organisms without also adopting the position that they
are part of a closed system of distinct economic units populating the
world economy – to carry the biological language further, if the nation-
state's boundaries are not imagined to be fairly impermeable, then such
nation-states' economic cytoplasm might leak out.

Although the Peace of Westphalia codified an understanding of space
as "a horizontal order of coexistent places that could be sharply delimited
and compartmentalized from each other" (Ó Tuathail 1996: 4), the view
of nation-states as containers for economies only really solidified during

the twentieth century. In particular, Mitchell (1998) argues that the collapse of the international gold standard in the early twentieth century marked a significant development in the representation of nation-states as powerful actors on the world stage, a representation tied indelibly to the idea that economies are nationally organized entities constituted and contained within national boundaries and that they subsequently interact with one another in the inter-*national* arena. This emergent notion was reinforced by the Keynesian macro-economic policies of the 1930s, in which the national scale was "discovered" (Claval 2008: 3) and in which it was the *national* economy that was theoretically and analytically privileged (Radice 1984). Hence, as they sought to make sense of the post-gold standard and Depression-era world in which they were living, economists and policy-makers increasingly came to focus upon entities such as "the national income," which became "an object of policy that could be scientifically predicted and manipulated by . . . the state's economic engineers" (Adelstein 1991: 171), courtesy of new econometric measures such as the "Gross National Product" and the "Gross Domestic Product," measures which themselves take the nation-state's boundaries as their spatial referents (Carson 1975). This notion became even more ingrained in much economic thinking once European decolonization began in earnest after World War II, given how entities such as "the British economy" or "the French economy" (with its overseas *départements* such as Réunion and Martinique legally a part of France) were often previously seen to be intimately tied up with those of their colonies. Whilst the collapse of the gold standard, the emergence of Keynesian economics, and decolonization were separate events, their combined effect was that in the mid-twentieth century the world increasingly came to be "pictured in the form of separate nation-states, with each state marking the boundary of a distinct economy" (Mitchell 1998: 90), such that nation-states could be reimagined as containers of "the economy" which, by definition, was conceived of as organized at a national scale. Despite the common belief that economies have long been imagined to be constituted in national terms, in fact this is a relatively recent discursive creation (Herod 2009).

In addition to being viewed as a container of the economic during the twentieth century, the nation-state has also been frequently imagined as a container of culture and politics, a container which has, it is argued, become (at least until recently) ever more powerful and hermetically sealed since 1648. This is particularly so in recent debates and writings about the supposed impact of globalization upon the nation-state, especially concerning whether the planet is currently experiencing processes of globalization or merely internationalization, a distinction which may seem somewhat enigmatic but is at the heart of how the nation-state – and hence the national scale – is conceptualized. Thus, in debates about contemporary changes in economic, political, and cultural life within the context of "globalization," many writers have argued that nation-states – and hence the national scale of economic organization – are still relatively powerful and spatially coherent actors on the world stage and that what has been happening in recent years is simply that more "things international" have been occurring between them. For instance, it may be the case that there is more trade taking place between, say, the UK and Japan, but if this is trade of goods made wholly within each of these two nations, then this is an exchange of commodities manufactured within economies that are primarily organized within national boundaries (Herod 2009). If, on the other hand, commodities are increasingly manufactured using components that come from several nations – as with a car assembled in the US but which has parts from the US, Canada, Mexico, and China – then the production line obviously transects national boundaries, an indication that economies are no longer contained within spatially discrete national territorial units.

Equally, in the political realm nation-states have been seen as containers of political identity at least since the nineteenth century. Hence, whereas previously there had often been a fairly loose association of certain political rights (such as those of citizenship) with national territories – in the case of the US, for instance, until the early twentieth century myriad states allowed non-citizens to vote in local, state, and even federal elections (Hayduk 2006) – "as nationalism grew during the nineteenth century, binding individuals to an imagined political community which had as its spatial expression the bounds of the nation-state, there was a growing convergence in many parts of the world between nation-states' physical territory, their citizenry, and various political rights" (Herod 2009: 55). In such a world, where political allegiance is still to particular territorial entities circumscribed by exclusive national boundaries, it is possible to have a degree of political inter-nationalism – as when governments or citizens join or support international organizations such as the United Nations – but there is no fundamental challenge to the supremacy of the nation-state as the circumscriber of political identity. On the other hand, in a world in which political identity may come to be aligned with other territorial units – as with moves to develop a sense of "European citizenship" in which residents of one EU country can vote in the elections of others, as is currently the case for elections to the European Parliament – then it is more difficult to sustain an argument that the national scale is the spatial resolution at which political citizenship is circumscribed. Similarly, in the cultural realm there is a long tradition of viewing "national cultures" (Frenchness, Brazilianness, etc.) as contained within the spatial confines of national boundaries – nation-states, in other words, have served as organizational boxes which have either been constructed around nations post hoc (as in the case of the German nation) or have been constructed to allow a sense of national identity to be developed within them after their construction (as in the case of the French nation). In a world experiencing internationalization, it is assumed that people's cultural identities remain fixed by the national scale, even if their nation-states participate in international organizations such as the UN – the people born and raised within the political boundaries of France feel French and have an identification with French culture, those born and raised within the political boundaries of Brazil feel Brazilian and have an identification with Brazilian culture, and so forth. On the other hand, in a world that is seen to be undergoing processes of globalization it is generally assumed that the nation-state's boundaries as geographical circumscribers of national culture are being undermined, such that "national cultures" seem less distinctive (one can find McDonald's restaurants in many different parts of the globe, for instance) and what is emerging in their stead is a kind of hybrid "global culture" involving "the fusion of . . . hitherto relatively distinct forms, styles, or identities . . . across national borders" (Kraidy 2005: 5).

Much of the debate over whether or not we have collectively entered an era of "globalization" rests on the status of the nation-state and whether it can be seen to still contain economic, political, and cultural life in the way in which it was assumed to do in the past (for more on this, see Herod 2009). There are a number of important issues that emerge out of this discussion with regard to how the national scale is conceptualized. First, questions concerning whether we are seeing processes of internationalization develop or whether they are actually processes of globalization only make sense if there is already an implicit assumption that nation-states are fairly hermetically sealed containers of politics and culture – it relies, in other words, on a particular construction of the nation-state and the national scale as fairly discrete entities whose spatial coherence is (or is not) being undermined by economic, political, and cultural processes which are weakening national boundaries' power to serve as territorial enclosers. Second, and relatedly, much of the rhetoric that globalization is undermining the nation-state as a "power container" (Giddens 1987: 172) relies for its potency on a narrative that nation-states and the national scale were much more significant in the past. However, it is debatable as to whether this was actually the case (Herod 2009, especially Chapter 7). Third, representing the national scale as, for instance, the container of national cultures that are now being hybridized by the weakening of national borders and the consequent mixing of domestic cultural practices, norms, and ideas with those from outside the nation-state assumes there is something that could be called "national culture" and that this evolves out of events taking place within national borders. However, not only is this a difficult position to sustain – can British culture really be thought to be the sum of the activities that have occurred historically within the geographical territory located off Continental Europe's northwest coast? – but it also raises the question of how the ideas of the nation-state and national culture have been used to discipline thought about culture: it is the concept of the national scale that allows differences between various regional cultures to be minimized (in the case of France, for instance, in such a narrative differences between Picardy, Gascony, Normandy, and Provence are minimized whilst those between France and Germany or Spain are emphasized). Finally, it is significant that even the choice of the word "hybrid" to describe the emergence of so-called "post-national" cultural identities contains within it the seed of a view of cultures as initially nationally defined, given that in biological terms a hybrid is usually created by the coming together of two quite distinctive biological entities from different *taxa* – the hybridization of national cultures, in other words, presupposes the existence of such national cultures (Box 5.5).

Feminist scholars have also explored how the national scale has been imagined as a container, often conceptualizing it in areal terms in much the same way as have their non-feminist colleagues. Hence, Mahon (2006: 457) has suggested that much feminist analysis of gender and social politics has "frequently conformed to the methodological national-ism that characterized mainstream research," largely "because welfare regimes, and more broadly gender regimes, had come to be consolidated at the national scale" – although she also accepts that "'[g]lobalization' is challenging this nation-centered assumption and, with it, the very way

---

Box 5.5 HYBRIDITY

The concept of hybridity emerged in the biological literature to refer to the crossing of two different parental types to produce an offspring, as when a male horse and a female donkey breed to produce a mule. The concept has been used by several post-colonial writers as part of an anti-essentialist discursive strategy for writing the history of colonialism. Perhaps most notable amongst these has been Homi Bhabha (1994), who has argued that during the age of European imperialism the contact between colonizer and colonized produced a hybrid society in which both were changed, a fact which helped to undermine the colonizer's power. Often, the concept assumes that two distinct "national" cultures interact — as when, for instance, the English conquered Ireland, with their interaction producing a hybridized Hiberno-English version of English in which Irish words have been incorporated into the English spoken in Ireland. However, in actuality, neither the Irish nor the English cultures were "pure," having been shaped historically by different influences beyond their boundaries (e.g., *via* trade or religious links), by various other conquering powers (e.g., the Vikings and the Normans, whose presence transformed in different ways the indigenous cultures of both England and Ireland relative to how they had existed before the Vikings and Normans arrived), and by each other prior to the English conquest of Ireland (e.g., through the migration back-and-forth between England and Ireland of travelers, who perhaps shared heroic ballads and stories which influenced the development and telling of each other's local legends). In the contemporary period, globalization has often been presented as a process in which the differences between distinct and pure "national cultures" are being erased and hybrid "global cultures" are taking their place, although this glosses over the fact that all "national cultures" have been shaped over the millennia by others to the point wherein understanding them to be "pure" is simply an ideological position.

Biologically speaking, it is possible to produce a hybrid by crossing different populations within a single species, which is often done in plant and animal breeding. However, in taking further the analogy of cultural hybridity, this understanding would not make much sense, as it would refer to instances of cultural mixing and innovation within a single culture qua species, rather than between cultures.

of doing feminist research." However, many have begun to challenge this approach to understanding the national scale as a so-called "objective scale" and to argue, instead, that "the national scale is produced through social and political processes that privilege particular identities and exclude others" (Silvey 2004: 492–93). Such recognition of the national scale's production as being highly gendered, Silvey (p. 494) contends, allows for better "conceptualization of [the] relational linkages between bodies, households and the transnational sphere." Likewise, Rose (1993) has argued that ideas of enclosure and the use of borders such as the national scale to do so represent a masculinist epistemological stance because such enclosures are about disciplining the – implicitly feminized – spaces contained within them. Thus, when the boundaries that would become, at independence, national borders were being laid down by various imperial powers in the Global South during the nineteenth century, the newly conquered lands were typically feminized through comparison with the female body (McClintock 1995: 24). As part of this feminization of space,

[a]s European men crossed the dangerous thresholds of their known worlds, they ritualistically feminized borders and boundaries. Female figures were planted like fetishes at the ambiguous points of contact, at the borders and orifices of the contest zone [and] women served as mediating and threshold figures by means of which men oriented themselves in space, as agents of power and agents of knowledge.

Indeed, images of the national space as feminized appear to be quite widespread at this time, as Goswami (2004: 199–203) shows in the case of nineteenth-century India. Paradoxically, McClintock posited, the establishment of boundaries and borders as feminized seemed to serve to counter a male loss of control over space – male explorers, soldiers, and colonial bureaucrats ritualized what they worried they could not really control.

Whilst McClintock has maintained that borders – and hence the national scale – have a history of being feminized, Mains (2002) has explored how the US Border Patrol has marshaled a series of masculinist tropes to "protect" the interior of the US by stopping illegal incursions into it, primarily from Mexico. Although the US heartland is presented in such tropes as feminized, vulnerable to moral corruption by foreign forces and thus "a space that requires protection" (p. 203), interestingly the national border itself is presented within the Patrol's narrative of its own activities as a highly masculinized place, one in which muscular policing strategies must be carried out if the sanctity of US territory and sovereignty are to be protected. In such tropes the idea of a distinct and clear boundary between "US culture" and the "Mexican culture" to be found just a few feet across the border both represents the unfolding of a masculinist epistemological stance and one in which the two nation-states are seen as containers of their respective national cultures. Such ideas that national borders as containers of social life must be policed and their robustness constantly reinforced rhetorically against external threats, however, are also quite common in other contexts where incursion by foreign bodies is feared. Hence, the national border has often been constructed rhetorically as an important prophylactic protecting the internal body politic from disease from abroad/outside, whether that disease is in the form of ideas (e.g., in the US strong, well-monitored borders have often been seen as necessary to protect against dangerous "foreign" ideas such as communism (Harris 2003)) or in the form of real disease, with this latter often used to mark as potentially dangerous and unclean immigrants seeking to cross the state's skin/border (Weibe 2009). Even the language of Cold War *realpolitik*, in which US policy was to "contain" communism so that it did not spread to infect more nation-states or the US itself, is emblematic of how nation-states were conceptualized in this discourse in box-like terms.

Much Marxist-inspired theorizing has also tended to view the national scale as a container of various social practices, particularly as these relate to the capitalist accumulation process. Thus, Smith (1984/1990: 173) has argued that scales such as the national scale allow events and people to be, quite literally, "contained in space." Such a position is similarly articulated by Harvey (1982: 404), who has maintained that the "territorial organization of the state – and the boundaries of the nation state are by far the most important – then becomes *the* geographical configuration *within which* [emphasis added] the dynamics of the investment process is worked out." Brewer (1990: 266) has also presented an areal understanding of the nation-state, suggesting that states have historically pushed out their boundaries "to the point where the resistance of other nation states limited further expansion" and that as "capital extends outside national boundaries" it creates links with other states. Likewise, Brenner (1998a) explored the tensions between capital's desire, on the one hand, to be mobile so as to seek out new investment opportunities elsewhere and its need, on the other, for fixity in space so that accumulation can take place, with this tension worked out geographically through the construction of various "space envelopes"/parcels of space (such as the national scale) wherein such parcels' interior spaces exhibit a certain coherence and share greater similarity to each other than to the interiors of other national spaces. Much of this writing owes its inspiration to French Marxist Henri Lefebvre, who conceptualized (1974/1991: 280) the nation-state as "based on a circumscribed territory" and as emerging out of the sixteenth-century push to develop a more integrated global economy in which, somewhat contradictorily, "the advent of a world market, implying a degree of unity at the level of the planet, [gave] rise to a fractioning of space – to proliferating nation states" (p. 351), whose territorial organization provided various segments of capital with the necessary spatial fixes to organize the forces of accumulation.

Summing up such understandings of the nation-state and national scale as container in the study of international relations (although their observations are also pertinent to other fields of study), Agnew and Corbridge (1995: 83–84) make three noteworthy points. First, they maintain that any reification of nation-states as "fixed units of sovereign space" serves to "dehistoricize and decontextualize processes of state formation and disintegration," as it presents states as timeless and placeless. Second, the frequent contrasting of the terms domestic and foreign/national and international and their presentation as polar opposites serves "to obscure the interaction between processes operating at different scales, for example, the link between the contemporary globalization of certain manufacturing industries and the localization of economic development policies." This phenomenon has been particularly common, they aver, in those international political economy approaches which have "fixat[ed] on the 'national' economy as the fundamental geographical entity." Finally, because the territorial state has often been "viewed as existing prior to and as a container of society . . . society [is viewed as an inherently] national phenomenon."

Globalization and networked states?

Whereas the idea of the national scale as a container of social life has a long history in various social sciences, including Geography, in more recent years an intellectual fascination with networks has impacted how nation-states and the national scale are conceptualized. In particular, several writers have suggested that we are seeing the emergence of various "post-national" states which are imagined to have quite different territorial configurations from traditional nation-states, a development which has implications for theorizing the national scale. Jessop (2008: 5), for instance, has argued that the concept of a "division between domestic and international affairs" only makes sense within the context of a Westphalian system of discrete nation-states in which the state "produce[s], naturalize[s], and manage[s] territorial space as a bounded container within which political power can be exercised." However, with globalization's alleged breaking down of the nation-state's territorial hegemony, several new modes of political organization may be emerging to challenge the position of the nation-state and the national scale, including "cross-border regional cooperation, a new medievalism, supranational blocs (e.g., the EU), a Western conglomerate state, and an embryonic world state." Part of this change in how nation-states and the national are conceptualized no doubt relates to material changes in the political economy of global capitalism (as with the emergence and strengthening of transnational political blocs such as the European Union) but some is also likely to result from changes in the metaphors that are used more broadly to understand the world, changes which undermine the representation of the nation-state as a fairly impermeable container of discrete economies, cultures, and politics – the rise of computer technology in particular has resulted in the language of networks becoming more prevalent when describing the world and no doubt this has impacted how nation-states are conceptualized.

The issue of the politics of representation is important when considering how the national scale has been conceptualized, for only with the withering of the power of the image of the nation-state as a container of discrete economies, cultures, and polities has it been possible to envision it in other forms. This is particularly important in the present context of globalization, given how the narrative promulgated by many proponents of the idea that globalization is undermining nation-states' power and threatening their sovereignty relies for its potency on the notion that nation-states were much more powerful in the past, propositions which are all subject to serious questioning (Herod 2009, especially Chapter 7). Thus, Appadurai (1996: 49) has suggested that there is an ongoing emergence of "postnational cartographies" and that nation-states, which could "once be seen as legitimate guarantors of the territorial organization of markets, livelihoods, identities, and histories," now appear to be "the only major players in the global scene that really need the idea of territorially based sovereignty," as many other social actors are "evolving nonstate forms of macropolitical organization: interest groups, social movements, and actually existing transnational loyalties." Despite the fact that there is much debate about how much, or even if, nation-states' power has actually diminished – Taylor (1994), for instance, has suggested that rather than states as containers withering away, they appear to have become a bit "leaky" in certain respects yet may actually have increased their control over territory in some other regards – the ubiquity of claims that a post-Westphalian system of weakened states is emerging seems to have opened the door to a more widespread imagining of nation-states as something other than containers.

Arguably, one of the most influential sets of arguments about the rise of networked social institutions has been that developed by Castells. In a series of publications (e.g., Castells 1996, 2004; Castells and Cardoso 2006) he suggested that although networks have always existed in human society, technological innovations are now making it possible for more organizations to evolve into a networked form of organization. Within this context, the idea of networked states has become increasingly popular with many writers. For instance, Christiansen (1996) has written about the emergence of a "networked polity" in the context of the EU, whilst the emergence of Hong Kong, Singapore, and Kuwait as important financial and industrial centers, Agnew and Corbridge (1995: 89) offer, is perhaps suggestive that "node and network" forms of political-economic organization may be emerging and strengthening. Likewise, Shen (2004) has suggested that the growing relationships between places such as Hong Kong and China or the Indonesia–Malaysia–Singapore growth triangle represent the emergence of networked "borderless region states." Paradoxically, perhaps, in some ways these political structures seem to mirror those of medieval Europe, where "there were few fixed boundaries between different political authorities [and r]egional networks of kinship and interpersonal affiliation left little scope for fixed territorial limits" (Agnew and Corbridge 1995: 84). Indeed, Anderson (1996) has even promulgated the rise of a "new medievalism" in this context, wherein political boundaries are flexible and sovereignty is not singular within particular territorial spaces but is multifaceted. This may be a somewhat ironic development, given that the Renaissance (at least in Italy) marked a period in which networks were increasingly made into nation-states as the "fluid urban factionalism" of the late medieval period gave way to "the birth of . . . regionally consolidated . . . state[s]" (Padgett and Ansell 1993: 1260). With the growth of cross-border linkages and the waning of the image of nation-states as almost hermetically sealed containers, it may be the case that some nation-states are now giving way to their precursor forms of governance.

However, others, such as Morgan (2007), have argued that what appears to be happening, at least in some nation-states such as the UK, is that rather than the emergence of a networked state we are witnessing the emergence of a "polycentric state," one which is bounded and porous, territorial and relational – that is to say, one in which political entities "come with no automatic promise of territorial or systematic integrity, since they are made through the spatiality of flow, juxtaposition, porosity and relational connectivity" (Amin 2004: 34). Likewise, Allen and Cochrane (2007) have suggested that political entities that have frequently been seen in terms of spatially continuous territorial units – they talk about regions, but the same could be said for nation-states – can be viewed, thanks to the emergence of more diffuse forms of governance, as becoming increasingly spatially discontinuous. Hence, they aver, regions [nation-states] "are being remade in ways that directly undermine the idea of a region [nation-state] as a meaningful territorial entity[, for] the governance of regions [nation-states] . . . now works through a looser, more negotiable, set of political arrangements that take their shape from the networks of relations that stretch across and beyond given regional [national] boundaries." The result is that the "agencies, the partnerships, the political intermediaries, and the associations and connections that bring them together, increasingly form 'regional' ['national'] spatial assemblages that are not exclusively regional [national], but bring together elements of central, regional and local institutions . . . that call into question the usefulness of continuing to represent regions [nation-states] politically as territorially fixed in any essential sense." Thus, in the case of the EU, the growth of "multilevel governance" (Aalberts 2002) and EU-wide inter-urban networks (Leitner et al. 2002) is resulting in both the growing geographical overlapping of political authority formerly held within national boundaries and the criss-crossing of those national boundaries by various social, cultural, and political networks.

Such empirical and discursive developments have not only led to a questioning of the territorial integrity and sovereignty of the nation-state and the national scale, but have also led to a more fundamental questioning of how national borders themselves have been conceptualized. For instance, Agnew (2008: 176) holds that whereas borders have typically been thought of as "reflecting an unambiguous sovereignty that ends/begins at a border or that must be overcome as such," a view that corresponds closely with the "national scale as container" trope, "border thinking should [instead] open up to consider (a) territorial spaces as 'dwelling' rather than national spaces and (b) political responsibility for pursuit of a decent life as extending beyond the borders of any particular state." For him, it is important to evaluate critically the way borders themselves are imagined because borders both "have real effects and because they trap thinking about and acting in the world in territorial terms. They not only limit movements of things, money, and people, but they also limit the exercise of intellect, imagination, and political will." Thus the challenge, Agnew contends, is "to think and then act beyond their present limitations." Hence, whereas from one viewpoint "borders are simple 'facts on the ground' (or, more radically, lines on the map) [which] exist for a variety of practical reasons and can be classified according to the purposes they serve and how they serve them," in the process enabling "a whole host of important political, social, and economic activities," from a more critical perspective they "are artefacts of dominant discursive processes that have led to the fencing off of chunks of territory and people from one another" (pp. 175–76). How borders are thought of has important implications for how space and territory themselves are conceptualized and represented.

For her part, Schlottmann (2008: 826) has examined how various spatial scales such as the national and the regional "are not only produced but also continuously performed in acts of communication." The key issue here, she has suggested, is how a language of borders or non-borders is made to perform (Butler 1990), regardless of whether borders themselves are becoming more or less hermetic sealers of national territory. Thus, Schlottmann maintains (2008: 824), "in order to grasp the resistance of closed spaces (both imagined and socially real) to their own dissolution, the analytical task posed is to examine their performance in everyday communication and to clarify the work of their hidden logics as well as their involvement in cultural, economic, and political practice." Thus, despite the recent "uncritical acceptance of fluids, folds, or networks as a pregiven, exclusive, and overriding postmodern [and post-Euclidean] ontology" (p. 825), and despite the fact that the idea of nation-states as containers "surely (over)simplifies contingency and complexity," the latter's embeddedness in the human consciousness means that such "closed spaces" are still likely to be viewed as "reasonable and [,] to a certain extent, indispensable tools for making the world intelligible by identifying, organizing, and structuring complex phenomena [, with the result that] instead of only searching for new and 'more adequate' spatial representations, the everyday use of the 'old' ones should also remain a subject of thorough sociogeographic inquiry" (p. 823).

SOVEREIGNTY MATTERS

In this final section I want briefly to explore the issue of sovereignty. In particular, much of the discussion about the weakening of the national scale as a spatial resolution has involved discussions of the reduction of nation-states' sovereignty in the face of various globalization processes. However, two issues in particular must be contemplated in this regard: i) there are clearly different types of sovereignty and different states may see these types impacted in quite divergent ways in the contemporary era; ii) the relationship between nation-states and emergent supranational entities such as the EU and the World Trade Organization is more complicated than expressed in a rather simplistic notion that the growing power of the latter automatically means a reduction in that of the nation-state.

Varieties of sovereignty

In much of the narrative about the alleged weakening of the nation-state's sovereignty and the undermining of the national scale of social life in the face of growing numbers of supranational organizations, sovereignty has often been assumed to be a unidimensional aspect of the nation-state: a nation-state either has it or it does not. However, as Krasner (1999) has pointed out, it is possible to distinguish at least four different sovereignty types: "recognition sovereignty" (the degree to which a state is accepted by others as part of the international community); "Westphalian sovereignty" (one nation-state's ability to not have its internal workings controlled by another); "border sovereignty" (a nation-state's capacity, ability, and willingness to regulate flows of information, commodities, money, and people across its borders); and "domestic sovereignty" (a nation-state's ability to determine and implement policies within its borders and beyond). Distinguishing between these types is important for considering whether, in fact, the national scale is being undermined by processes such as globalization. For instance, it is quite possible that a nation-state may see an erosion of one type of sovereignty but an augmentation of another. Hence, the coming to power of a military government in a coup may result in lost recognition sovereignty but a clamping down on flows of information and other things crossing its borders (i.e., an enhancement of border sovereignty). Likewise, a nation-state's membership in a supranational entity such as the EU may reduce its border sovereignty but boost its recognition sovereignty.

Several important points come out of this discussion. First, different nation-states clearly may enjoy different types of sovereignty. Second, depending on the specific nation-state involved, these various aspects of sovereignty may reinforce one another or they may pull in opposite directions. Hence, some states may be strong in all four dimensions whereas others may have strong recognition and Westphalian sovereignty but have difficulty with, say, border sovereignty – as with the United States, which is generally accepted as a legitimate member of the inter-national community and does not have its internal workings controlled by other states but which has trouble securing its border with Mexico. Third, it is important to bear in mind that a particular nation-state's configuration of these different types of sovereignty may vary historically, such that its recognition or domestic sovereignty may be strong at one time but weaker at another. Fourth, the part should not be mistaken for the whole with regard to the apparent undermining of national borders. Hence Newman and Paasi (1998: 199) argue that much of the rhetoric that borders are becoming less significant, and thus that the national scale is less important, "has its roots in the [English-speaking] western European territorial narrative [and is] propagated by scholars within the western European and North American tradition," where the rise of entities such as the EU and the North American Free Trade Agreement zone mean that boundaries appear less important than previously. The result, they sustain, is that many authors "perceive their own contexts as constituting 'The Texts' of the current world," even though there are few parallels across the globe to the gradual opening of boundaries in Western Europe and, moreover, places such as the former Yugoslavia and Soviet Union have actually witnessed the establishing of new national boundaries in recent years between what were previously sub-nationally constituted republics. "The disappearance of boundaries and all that it entails," they suggest, is therefore "relevant to only a small part of humankind. For the rest, territorial partitioning seems to remain the order of the day."

Finally, although much writing has assumed that reductions in the nation-state's sovereignty represent an undermining of the national scale, it is important to distinguish between a nation-state's sovereignty and its ability to exercise power. In this regard, Elden (2005) draws on the historical account of German philosopher Gottfried Leibniz who, having been asked by the Duke of Hanover to clarify rulers' position within the Holy Roman Empire during the late seventeenth century, distinguished between, on the one hand, the power to demand obedience and loyalty, and, on the other, sovereignty, which Leibniz viewed as concerned with territory. Similarly, Agnew (2005: 437) has maintained that "sovereignty is neither inherently territorial nor is it exclusively organized on a state-by-state basis." To explore this he puts forward a sovereignty model that is different from the one that typically assumes sovereignty and state authority are two sides of the same coin. Contending that "effective sovereignty is not necessarily predicated on and defined by the strict and fixed territorial boundaries of individual states" (p. 438), Agnew argues for a schema which takes account of sovereignty's social construction, its association with hierarchical subordination, and its deployment in territorial and non-territorial forms. On this basis he identifies four "sovereignty regimes" which, he contends, result from "distinctive combinations of central state authority (legitimate despotic power) on the one hand, and degree of political territoriality (the administration of infrastructural power) on the other" (p. 437) (Box 5.6).

First, there is what he calls the "classic example" of sovereignty, in which the nation-state still has a high degree of effective central political authority and in which its "despotic and infrastructural" power is still largely deployed within a bounded state territory, even if it is increasingly dependent on economic interchange with the outside world. The idea of a Westphalian system of states largely draws upon this conception. Second, there is the "imperialist example," which is the "exact opposite" situation as

[n]ot only is central state authority seriously in question because of external dependence and manipulation [but] state territoriality is . . .

Box 5.6 DESPOTIC POWER AND INFRASTRUCTURAL POWER

In developing his argument about sovereignty regimes, Agnew draws upon Michael Mann's (1984: 188) work, specifically his distinction between "Despotic Power" and "Infrastructural Power." Mann defines these two terms as follows: Despotic Power "denotes power by the state elite itself over civil society," whereas Infrastructural Power "denotes the power of the state to penetrate and centrally co-ordinate the activities of civil society through its own infrastructure."

---

subject to separatist threats, local insurgencies, and poor infrastructural integration[, i]nfrastructural power is weak or nonexistent, and despotic power is often effectively in outside hands (including international institutions such as the World Bank as well as distant but more powerful states).

Third is the "integrative example," as represented by the EU, in which "sovereignty has complexities relating to the coexistence between different levels or tiers of government and the distinctive functional areas that are represented differentially across the different levels, from EU-wide to the national-state and subnational-regional." In such a situation "the territorial character of some of its infrastructural power is difficult to deny" (as with, say, the Common Agricultural Policy, which applies across the EU), though central state authority for both the entire supranational organization and the member states may be "weaker than when each of the states was an independent entity." Finally, he presents the "globalist example," of which the best current example is probably the effective sovereignty exercised by the US "within and beyond its nominal national boundaries and through international institutions within which it is particularly influential (such as the IMF)" and into which it has attempted to recruit other states, "by cooptation and assent as much as by coercion" (pp. 445–46). What is key for Agnew is that these different types of sovereignty have been dominant in different parts of the globe and at different time periods, a reality which provides for a much more complex historical geography of sovereignty's waxing and waning and hence of the coherence of the national scale.

Globalization, supranationalism, and the national

Arguably, if any one force or process could be seen to be undermining the national scale's coherence in the contemporary period it would undoubtedly be that of globalization, itself typically conceived of as a phenomenon in which transnational flows of capital, together with the growth of supranational entities such as the EU, are heralding the "twilight of sovereignty" (Wriston 1992) and the "end of the nation-state" (Ohmae 1995). However, here, too, it is important to examine critically the argument that globalization and the growth of supranationalism necessarily challenge the power of the nation-state and, hence, the national scale. Specifically, there are five issues to consider.

First, although much of the narrative about globalization's effect on the nation-state has suggested that globalization is weakening the nation-state in *toto*, in many instances we have seen an increase in the nation-state's powers. In the case of the EU, for instance, although internal barriers to migration have been much reduced in the past two decades, barriers with the outside world have been much increased, leading some to proclaim the emergence of a "Fortress Europe." Likewise, whereas the power of nation-states to regulate flows of information and capital across their borders seems to have diminished, their ability to regulate workers and welfare recipients appears to have increased in recent years – as Peck (2001) has pointed out, for instance, welfare "reform" has largely involved simply making it more difficult for the unemployed to access benefits.

Second, membership in supranational organizations may actually augment nation-states' powers. Hence, Milward (1992) has argued that the EU's creation, rather than undermining many Western European nation-states, actually facilitated their resurrection in the post-1945 era, allowing them to use the nascent European community as a protector against Germany. More recently, Hudson (2000) has suggested that by providing a trading bloc to rival the US, the EU has allowed its constituent member states to augment their economic power in a way not possible were they still each acting on their own. Equally, some relatively weak nation-states – say, Mali – may find that being members of a supranational organization such as the World Trade Organization (WTO) strengthens their bargaining position vis-à-vis more powerful states because now all states have to play by the same set of rules. Hence, should a more powerful state seek unfairly to pressure Mali, the latter would be able to seek support from the broader WTO to ensure it is protected by the organization's rules. Third, rather than a totalizing discourse which supposes that globalization and its trappings are weakening the (singular) nation-state, it is important to recognize that different states are likely to be impacted in different ways, based upon their power resources, history, how they are networked into the broader global economy, and so forth. Put another way, the US's capacity to resist having its sovereignty and power undermined is likely to be much greater than that of myriad other countries.

Fourth, the language of nation-states having "more" or "less" sovereignty within a globalizing economy fails to recognize that, rather than there being either an undermining or augmentation of sovereignty, we may be seeing instead a more complex and interscalar redistribution and reworking of power between and within different levels of government – rather than an absolute loss of power, it may be that power is simply recoalescing at the level of the local state. Yet, even if this is the situation, given that the nation-state draws much of its power from the strength of the local state, the nation-state may actually have just as much power as previously, only now it is the case that if we are to analyze this power we must focus upon the local state. Moreover, such a narrative – of power being relocated from the national to the local state – only makes sense if the national scale and other scales are seen in areal terms as separate entities. However, if scales are conceptualized instead in terms of a network or as part of a highly integrated nestedness in which a "multifaceted causality runs in virtually all directions among the different levels of society" (Hollingsworth 1998: 28), then it is much more difficult to envision a relocation of power from one distinct scalar level to another. Finally, this rhetoric suggests that globalization and supranationalism are impacting a passive nation-state. However, such a narrative conveniently ignores the fact that it has been nation-states that have often facilitated globalization and supranationalism – thus, it is nation-states that relax regulations to allow cross-border flows of capital, goods, and people and it is nation-states that make the decision to join supranational organizations such as the EU or WTO. Given this, it is much more difficult to sustain the argument that the national scale is being diminished by globalization and/or supranationalism, as it is national-level politicians who must agree to such legislation – only nation-states can join the EU, for instance.

CONCLUDING REMARKS

In this chapter I have explored several issues concerning the national scale, including how it has been seen to coincide with the territorial boundaries of the nation-state, how nation-states have been imagined historically (as living organisms, for instance), how the image of the Westphalian state encourages an areal view of the national scale but that other views (such as those associated with networks) have become more popular in recent years, and how some of the arguments suggesting that globalization and supranationalism are undermining the national scale bear close scrutiny. An important aspect to bear in mind here is that the way in which many concepts – sovereignty, territoriality, nation-state, national scale – are conceptualized has significant bearing upon how others are understood. Hence, for instance, Agnew and Corbridge (1995: 84) have maintained that only with fairly fixed national boundaries does the concept of sovereignty make much sense. Thus, during the Middle Ages, when the key question was where an individual stood within a particular hierarchy – peasant, noble, king – so that they could figure out to whom they owed loyalty and who (if anyone) owed them loyalty, the concept of territorial sovereignty did not mean much. Rather, communities were "united only by allegiance and personal obligation rather than by abstract conceptions of individual equality or citizenship in a geographically circumscribed territory [and space] was organized concentrically around many centres depending upon current political affiliations." The lesson to be learned here is that arguments that the national scale is being undermined are only logically coherent if they are built on a whole series of other assumptions – that the national scale is an entity discrete from other scales, that the national scale serves to enclose all "sub-national" activities and the territory of the nation-state, that the national scale is the scale at which sovereignty is exercised, that the national scale is inherently subservient to scales that are "larger" than it (such as the global), and so forth. Once these supporting assumptions are challenged, the more difficult is it to sustain the initial argument that the national scale is withering away.

It is also important to bear in mind that although an areal view of scale sees the national scale as circumscribing particular nation-states' territories, such that the national scale delineates where one nation-state ends and another begins, within such territories various organs of the national-level state operate at sub-national levels – in other words, not all "national" activity is about dealing with other nation-states. In the case of France, for instance, this sub-national focus has historically been formalized through the appointment by the national government in Paris of a prefect to represent the Ministry of the Interior on local governments throughout France, with the prefect serving (until recently) as the chief executive of the government of each local *département*. In other cases, although there may not formally be a representative of the national government on local governments, the representatives of the national state may play significant roles at the sub-national scale – building roads, managing land, and so forth. The point is that although the national scale is often seen to be coincident with the spatial boundaries of the nation-state, the nation-state itself does not only operate at a national scale.

Lastly, there is one final matter of the relationship between the national scale and others. Generally, much of the writing about the development of the nation-state and the securing of a national scale has assumed that modern nations have been "built from political centers outward and imposed upon marginal groups or peripheral regions in a process of cultural and institutional 'assimilation' and 'integration'" (Sahlins 1989: 7–8). In such a view, power is seen to have emanated from some capital city in a process in which the scales of local and regional life and identity were subdued by the national scale in much the same way that the global scale is now seen to be emasculating the national scale. In such a representation there is clearly a one-way relationship between the national and sub-national scales, wherein the former colonizes and dominates the latter, such that local and regional identities and power structures are replaced by national ones. The development of "[n]ational identity" in such a narrative "means replacing a sense of local territory by love of national territory." Sahlins, however, has maintained that rather than the process of crafting such a national scale being one in which sub-national scales are colonized by scales that are "larger" or "higher" than they, such that in the case, say, of France (his focus) "peasants bec[a]me national citizens only when they abandon[ed] their identity as peasants [and] a local sense of place and a local identity centered on the village or valley [were] superseded and replaced by a sense of belonging to a more extended territory or nation," in fact "state formation and nation building were two-way processes [and] local society was a motive force in the formation and consolidation of nationhood and the territorial state." There was a "dialectic of local and national interests" which "produced the boundaries of national territory" – the national scale, in other words, did not appear on the world stage fully formed but had to be actively created in dialectical tension with other scales. Such a finding raises important questions for our own age, when the power of the global is seen to be growing at the expense of the national. Hence, rather than globalization necessarily placing the global scale at odds with the national in a kind of zero-sum game, the connectivity between them means that the national and the global mutually shape how each other is constituted.

6

THE GLOBAL

Globalization [is] the rise of supraterritoriality.

Scholte (1996: 565)

The two extremes, local and global, are much less interesting than the intermediary arrangements that we are calling networks.

Latour (1993: 122)

The question is . . . no longer one of the *global* versus the *local*, or of the *transnational* versus the *national*. It is, first and foremost, a question of the sudden temporal switch in which not only inside and outside disappear, the expanse of the political territory, but also the before and after of its duration, of its history; all that remains is a *real instant* over which, in the end, no one has any control.

Virilio (1997: 18)

As globalization has become the leitmotif of the past three decades or so, the global has come to be viewed by many as the scale to which all social actions must respond. This is evident in the myriad statements that take the form of "we must do X *because* of the power of global capital and global institutions." In such discourses the global has become the scale from which there is no possibility of escape, "the scale which denies us the geographical option of fleeing to a distant spatial sanctuary in some remote corner of the planet wherein we can insulate ourselves from the consequences of contemporary social and natural processes" (Herod

2009: ix). Whether presented in terms of the delocalization of social life or the loss of national sovereignty, the global seems to have emerged as the *primus inter pares* scale of late twentieth-/early twenty-first-century social life and social theorizing.

Given this, here I examine three significant aspects of the global scale as manifested materially and ideationally. First, I outline some of the ways in which the global scale has been constructed historically. In particular, I explore how material practices of building a global scale of human existence have been reflected in ideas about the global and how, in turn, such ideas have shaped how the global as a material scale has been constructed. Second, I explore how the global scale has been understood relationally, particularly in connection with, and in opposition to, its Other – the local. Finally, I consider some of the literature on "global cities," which have often been presented as nodes where local and global processes meet in the form of a new type of planetarily linked urbanization.

CONSTRUCTING THE GLOBAL

Although colonialism and the growing transnationalization of corporate organization mean that the global has become an increasingly important scale of social and economic organization during the past five centuries or so, the global scale itself has longer held an important place in the thoughts of writers of various political persuasions, with some of the earliest imaginings deriving from the ancient world. Cosgrove (2001, 2003), for instance, has suggested that a global consciousness can clearly be identified as existing in the pre-Christian era. Thus, the idea that local attachments to family, community, and ethnic background should be sublimated to a sense of global unity had great significance for Greek and Roman Stoic philosophers and, later, for Christian theology. Such views were undoubtedly reinforced by the discovery that the Earth is round, with credit for so illustrating the planet's shape usually given to sixth century BCE philosopher Thales of Miletus. A focus on the global was also reflected in medieval Christian iconography, which used circles (as with halos) and spheres to emulate divine perfection, with Christ frequently portrayed holding or standing upon a globe (Box 6.1). Equally, despite the popular (but inaccurate) belief that they saw the Earth as flat, early Christian scholars – following Claudius Ptolemy's *Guide to Geography* – argued that there was a great equatorial ocean separating the landmasses of the northern and southern hemispheres and that such hemispherically balanced landmasses illustrated the harmony in God's world and the planet's geographical unity. Such iconography continued into the nineteenth century, as various missionary societies frequently used images of the globe (Cosgrove 1994). More recently, such a theological focus on the global has been highlighted by Goudzwaard's (2001: 20) argument that the Christian church was always intended to be a global community and that "God's economy entails its own style of globalization, oriented to the coming of his Messiah King." At the same time, however, it is important to note that some other Christian groups, whilst recognizing the global as a key scale, have done so from a quite different perspective, viewing globalism as something to be avoided and/or as the harbinger of calamity. Hence, many fundamentalist groups, drawing inspiration from the Book of Revelation, in which Satan is seen to orchestrate global domination through establishing a worldwide government, have perceived the global as a scale to be avoided.

If many early ideas about the global were shaped by theology, developments in the early modern period in astronomy, natural history, cartography, and philosophy, together with the increasing spatial integration of an emergent capitalist world-system (Wallerstein 1974) fostered by European colonialism, both facilitated the growing material construction of a global scale of social organization and further encouraged a global outlook. Kirby (2002), for instance, suggested that in Europe developments in astronomy – themselves facilitated by technical improvements in glass-making – encouraged writers to consider the Earth in global terms, especially with regard to other spheres in the solar system (the Sun and planets). For her part, Pratt (1992) has argued that European exploration of Africa, Asia, and the Americas, together with Magellan's 1519–22 circumnavigation of the globe, the mapping of the continents' coastlines, and the development of the trans-Atlantic slave trade, with its resultant fusion of the economies of Africa, the Americas, and Europe, encouraged a rising "planetary consciousness" from the sixteenth century onwards. The result, she averred, was the development of a "European global . . . subject" (p. 30) and the "construction of global-scale meaning" (p. 15). Equally, the German philosopher Immanuel Kant argued that the Earth should be seen as a globally integrated whole and that, because its spherical form meant that there were no actual "ends of the Earth" to which individuals might go to escape their fellow humans, people were forced to "support one another,

Box 6.1 THE GLOBE AND CHRISTIAN ICONOGRAPHY

The image of the globe has frequently been used in Christian iconography, dating to the days of the early church. This is because circles and spheres are seen as appropriate symbols of divinity, for they have neither beginning nor end. Furthermore, the idea that Christ is the *Salvator Mundi* ("Savior of the World") encouraged artists to use images of the globe in their work, as in the painting in Figure 6.1 of Christ holding the Earth in his hand. As a result of such ideas, the Christian *globus cruciger*, an orb (from the Latin *globus*) with a cross (Latin: *crux*) atop it, which represents Christ's dominion over the world, has long been a symbol of religious and royal authority.

Such associations between Christ and the globe have continued into the present day. Hence, when the famous statue of Christ the Redeemer ("O Cristo Redentor") which overlooks Rio de Janeiro was being planned, one design considered included Christ holding a globe. However, this was eventually dropped in favor of the current statue of Christ with wide-open arms.

Figure 6.1
Christ with globe – "Salvator Mundi," Lower Rhine, 1537–45 (oil on panel) by German School (sixteenth century), Deutsches Historisches Museum, Berlin, Germany/© DHM/ The Bridgeman Art Library/ Nationality/ copyright status: German/out of copyright.

since they [are] not able to disperse themselves to infinity" (quoted in Cosgrove 2001: 176). This growing fascination with the global is reflected in, amongst other things, language – the *Oxford English Dictionary*, for instance, dates the first usage of the word "global" to 1676.

Arguably, one of the most significant developments in this regard was the growing knowledge that Europeans began to gain of Australia, following their early seventeenth-century "discovery" of the great Southern Continent (*terra australis*). Although Australia soon proved not to be the southern landmass hypothesized by either Ptolemy or early Christian theology, the confirmation of such a large landmass in the southern hemisphere did seem to attest that the Earth was in some kind of global equilibrium. Moreover, the continent quickly became a canvas upon which European writers could paint large their fantasies about themselves, as Australia became Other to Europe in a globally integrated system of binaries. Thus, Australia's inhabitants were viewed quite literally to be antipodean, as they were imagined to have feet where their heads should be. Indeed, Australia seemed opposite to Europe in innumerable ways – as the writer John Martin (quoted in Jack 1988: 61) put it in the 1830s: "trees retained their leaves and shed their bark instead; the more frequent the trees the more sterile soil. The birds did not sing, the swans were black and the eagles white. The bees were stingless, some mammals had pockets while others laid eggs. It was warmest on the hills, coolest in the valleys." With Australia and Europe imagined as opposites within a global whole, European fascination with binaries concerning reason and anti-reason, normalcy and abnormalcy, progress and backwardness, and rationality and irrationality were played out spatially: "Europe served as the place of reason, normalcy, and rationality, whilst Australia was its opposite and Other, a place requiring disciplining through the introduction of European ways of thinking and living" (Herod 2009: 37).

Europeans' geographic and scientific discoveries, within the context of a developing Enlightenment sense of rationality and order, profoundly shaped how they came to consider the globe. Using an approach informed by feminist, Marxist, and post-colonial theory, Pratt (1992: 30) has argued that the "European global subject" that developed from European exploration of the world beyond its shores increasingly came to be viewed by eighteenth-century writers such as Daniel Defoe as "male, secular, and lettered," with the latter descriptor referring to the fact that European naturalists' journeys to strange foreign lands to catalogue flora and fauna asserted "powerfully the authority of print, and thus of the class which controlled it." Such scientific writings, she averred, crystallized "global imaginings of a sort rather different from the older navigational ones," which had filled in "the blank spaces of maps with iconic drawings of regional curiosities and dangers – Amazons in the Amazon, cannibals in the Caribbean, camels in the Sahara, elephants in India." Whereas early views had adopted an approach which saw newly discovered things somewhat in their own terms, as individual types, Pratt argued that eighteenth- and nineteenth-century European efforts to map out the globe's natural and human environments using the cataloguing systems developed by taxonomists such as Carl Linnaeus "conceived of the world as a chaos out of which the scientist *produced* an order." The result was that "[o]ne by one the planet's life forms were . . . drawn out of the tangled threads of their life surroundings and rewoven into European-based patterns of global unity and order," such that the "lettered, male, European . . . eye that held the system could familiarize ('naturalize') new sites/ sights immediately upon contact, by incorporating them into the language of the system" (p. 31). The global whole, in other words, increasingly became the point of reference for all questions concerning social or natural order and an object's place in that order – a fact brought home when the discovery of Australian animals such as the duck-billed platypus forced Europeans to tweak their global classificatory systems. Moreover, because scientists – with few exceptions, men – were the discipliners of this knowledge, the global was implicitly considered to be male.

Such an interest in the global continued in the nineteenth century as the planet was progressively more knitted together through imperialism and the geographical spread of capitalism. In particular, as transoceanic journeys became more common, the need to develop a global system of time coordination became ever greater. The origins of such dated back to the first European circumnavigations of the globe, when mariners had noticed that, depending upon which way they traveled, they might gain or lose a day relative to the time in their home ports. Thus, sailors who ventured westwards from Europe into the Pacific (principally the Spanish, who arrived *via* the Americas) discovered that their calculation of the date was one day behind those who had come eastward via the Cape of Good Hope (principally the Portuguese, the Dutch, the French, and the British) – a Spanish Wednesday was a Portuguese Thursday. These disparities led to growing calls for some sort of global time standard, calls which naturally raised the question of setting a Prime Meridian and its Other, an International Dateline. Certainly, efforts to institute a Prime Meridian by which to measure longitude – and thus to determine the relative time at any point of the Earth's surface at any given moment – stretch back at least to the ancient Greeks. Thus, the Greek astronomer and geographer Hipparchus proposed a meridian that passed through Rhodes, whilst Ptolemy suggested that the Canary Islands, considered the westernmost location of the known world, should serve the purpose. However, by the nineteenth century the growth of sea voyaging had made instituting an internationally recognized Prime Meridian more necessary than ever, even as the growth of nationalism made choosing an appropriate longitude more contentious. Thus, many Americans argued that the National Observatory in Washington, DC, should serve as a global Prime Meridian, whereas many Britons asserted that the Royal Observatory at Greenwich should do so, whilst the French and Spanish respectively argued for Paris and Cadiz.

Efforts to determine a Prime Meridian and an International Date Line both reflected and reinforced evolving views of a global perspective on time and space in the nineteenth century. They were closely connected with attempts to develop more broadly a global system of time and space coordination through the initiation of a planetarywide scheme of standardized time zones, with such an initiation itself largely reflective of the increasing influence in everyday life of two technological innovations: the railroad and the telegraph. Thus, although the development of relatively accurate clocks had allowed people traveling east or west to recognize that local times differed from the times in the cities from whence they were journeying, and although there had been attempts to develop standardized time zones in the eighteenth century (in 1792 London adopted "local mean time" to replace the city's myriad times derived from the Sun's movement across the sky), it was not until the spread of the railroads in the 1840s that much urgency was given to endeavors to standardize time, as trains operating under different times on the same tracks could prove decidedly hazardous. Consequently, in 1848 Britain's railways adopted Greenwich Mean Time (GMT) as their national time, an act which helped both improve railroad safety and reinforce notions of national unity, even as many towns continued to observe local time and parliament did not designate GMT as mainland Britain's official time until 1880. Across the Atlantic, in 1849 most New England railroads adopted a single regional time, though two 1853 accidents, when trains operating according to different time reckonings collided, quickly led the rest also to do so (Shaw 1978). As train tracks increasingly tied different parts of the US together, the railroad companies moved toward a national system of time, adopting standardized times across North America in 1883. Simultaneously, developments in the telegraph provided instant communication across space, although at a price as many astronomers charged fees for time services based out of their observatories (Bartky 2000) – market forces, in other words, played an important role in spreading standardized time's sway over various regions. These developments encouraged people to think more globally about time and space, in two ways. First, the telegraph allowed people in places physically distant from cities such as London or Paris to know what was the time in such metropoles, as an electric current could be sent down the cables at specific times (such as on the hour) to cause a bell at the other end to clang. Second, the building of international cable networks promoted a growing sense of global simultaneity, as news of events in distant places could be more rapidly transmitted across space.

The laying down of cable networks across the planet was central in the nineteenth-century intensification of a sense of globality. The first international cable to be of any commercial success was that established between Dover, England, and Calais, France, in 1850–51 (White 1942). This was quickly followed by a transatlantic cable – described by the US Minister to France as the "umbilical cord with which the old world is reunited to its transatlantic offspring" (cited in Weber 1992: 105). The sense of a budding global outlook which this engendered is quite evident in contemporary reports. Thus, the editors of the London *Times* remarked that "[d]istance as a ground of uncertainty will [now] be eliminated from the calculation of the statesman and the merchant," for as the

[d]istance between Canada and England is annihilated [and] the Atlantic is dried up . . . we become in reality as well as in wish one country. . . . To the ties of a common blood, language, and religion, to the intimate association in business and a complete sympathy on so many subjects, is now added the faculty of instantaneous communication, which must give to all these tendencies to unity an intensity which they never before could possess.

(*Times* August 6, 1858: 8)

The feeling of global simultaneity was reinforced in the 1860s when a second, more reliable, connection was put down and the Atlantic became viewed as "so little impediment to conversation" that it was "possible ... [to hold a] conversation at the same time with Europe and America ... [and for] all the civilized world to communicate instantaneously" (*Times* September 21, 1866: 6). The greater connectivity facilitated by the cable, the *Times*'s editors felt, would encourage global cultural homogeneity and understanding, as the new technology meant that "America cannot fail to live more in Europe and Europe in America [such that f]or the purposes of mutual intercourse the whole world is fast becoming one vast city" (*Times* July 30, 1866: 8). Indeed, the "three old Continents" were now increasingly considered by many "a single mass" (*Times* July 27, 1866: 9). Cables were also completed linking London with India (1864) and Australia (1872), with the latter cutting the time to communicate between London and Sydney from 45 days to 24 hours, a feat which greatly intensified feelings of global connectedness. By 1880 some 90,000 miles of cable had been laid, mostly by the British, who were keen to link together their far-flung empire.

The cable, however, was not only important for transmitting information more readily across great distances but it was also central in creating what French naval officer Octave de Bernardières called "an immense geodesic network that would encompass the entire globe, fixing precisely its form and dimensions" (quoted in Galison 2003: 141). Hence, as the cable stations established across the planet by the Great Powers allowed ever more accurate measures of time to be broadcast globally along the telegraph lines which were increasingly forming telegraphy's "wireless girdles around the world" (Andersen 1913: xiii), an ever more accurate mapping of territory could be undertaken. This allowed for greater control of such territory, and many expeditions were sent off to establish stations – a central element in France's integration of Senegal into its empire, for instance, was the French Bureau of Longitude's decision to send scientists to Dakar to determine the port's exact latitude and longitude, for this would not only allow French cartographers to map out more accurately their colony's interior and so help put down resistance but also facilitate French efforts to extend their cable system along the West African coast down to the Cape of Good Hope. The criss-crossing of the globe with telegraph cables quite literally bound every continent together into an increasingly interconnected system of global communication, even as various powers sought to turn such a system to their advantage. Indeed, as Galison (2003: 175) has commented, in the struggle for imperial domination "[l]ongitude, train tracks, telegraphy, and time-synchronization reinforced each other . . . [for e]ach showed a different facet of a new global grid."

One of the most intriguing, perhaps, indicators of the sense of the global that had emerged by the late nineteenth century was the 1879 suggestion by Canadian railroad promoter Sandford Fleming to establish a singular world "Cosmopolitan Time" wherein every clock on Earth would simultaneously display the same time (Blaise 2000). Whilst mocked by many, Fleming's proposal nevertheless reflected a growing concern that some kind of global time standardization needed to be developed. Although there had been some earlier efforts to do so – Italian mathematician Giuseppe Barilli (who went under the pseudonym Quirico Filopanti), for instance, had argued for such a system in his 1858 book *Miranda!* – it was not until 1884, when 41 delegates representing 25 countries met in Washington, DC, to decide upon a mechanism for coordinating time on a global scale, that much progress was made. The International Meridian Conference produced several outcomes, including that the Prime Meridian should run through Greenwich, England, and that there should be a universal day of 24-hours' length beginning at midnight, Greenwich time. Significantly, the delegates decided that their decisions should not interfere with the use of local times in various countries, with the result that the time zones were actually to be adopted through countries' internal legislative processes. Hence, in the case of the US, although five zones were delineated in the 1880s by the railroads, with such times subsequently adopted by most large cities, it was not until 1918 that Congress formally adopted the Standard Time Act. Other countries likewise began adopting legislation to bring their local times into correspondence with the schemes proposed by the conference, such that by the late 1920s most countries had done so and a global sense of time – and hence of space, given that time is measured by one's distance east or west of Greenwich – had been more fully assumed across the planet.

If developments in telegraphy and colonial conquest provided the material foundation for a global system of time–space coordination, novelists increasingly sought to ride the global spirit of the times that this engendered. In the francophone world, the leading writer in this regard was Jules Verne, who produced a series of works exploring the new global spatial and temporal horizons then being opened, including *The English at the North Pole* (1864), *Journey to the Center of the Earth* (1864), *From* the Earth to the Moon (1865), Twenty Thousand Leagues Under the Sea (1869–70), and Master of the World (1904). Arguably, though, it was Verne's Around the World in Eighty Days (1872/2004) which most encapsulated the age's new weltanschauung ("world view") (Box 6.2). In the English-speaking world, H.G. Wells likewise explored issues of globality in several novels, including A Modern Utopia (1905), wherein he imagined the formation of a world state that would bring about social reform, and The War of the Worlds (1898), in which he delved into the topic of global-scale inter-planetary conflict but also played on the image of scale-shifting in a fashion similar to how William Blake had done in his 1803 poem Auguries of Innocence. Thus, whilst Blake had written of seeing a "World in a Grain of Sand," Wells (1898/2006: 4) remarked in his novel's opening para-graph that whereas "[w]ith infinite complacency men went to and fro over this globe about their little affairs, serene in their assurance of their empire over matter," it was "possible that the infusoria [microorganisms] under the microscope do the same," a formulation designed to suggest humans' diminutiveness as viewed from an extra-global perspective. US writer Frank Norris (1901/1967) similarly highlighted a fascination with the global in his novels concerning how the telegraph had linked California wheat growers' actions into events in India, Russia, and Argentina, whilst German film director F.W. Murnau's 1922 production of the vampire movie Nosferatu also addressed issues of globality and the global spread of capitalism – the Transylvanian vampire invests in the German property market, for instance.

Within the world of political writing, too, the global scale came to be seen as increasingly significant, especially as its creation was related to the geographical spread of capitalism and imperialism. Hence, Marx and Engels (1848/1948: 12) posited that capitalism's expansionary nature meant that it would eventually spread out "over the entire surface of the globe [and] nestle everywhere, settle everywhere, establish con-nections everywhere." Indeed, such was the globalizing power of capital, they averred (p. 13), that the

> bourgeoisie, by the rapid improvement of all instruments of production, by the immensely facilitated means of communication, draws all nations, even the most barbarian, into civilization. The cheap prices of its commodities are the heavy artillery with which it batters down all Chinese walls, with which it forces the barbarians' intensely obstinate hatred of foreigners to capitulate. It compels all

nations, on pain of extinction, to adopt the bourgeois mode of production; it compels them to introduce what it calls civilization into their midst, i.e., to become bourgeois themselves. In one word, it creates a world after its own image.

Thus, it is no accident that Marx (1867/1976) ended the first volume of *Capital*, in which he detailed the internal workings of capitalist accumulation and the production of surplus value, with a chapter on colonization and how British capitalists exported surplus labor and capital to the colonies (for a more developed theoretical exegesis, see Harvey 1985). At the same time, Marx and Engels also argued that the opposition to capitalism would likewise have to be global in nature, epitomized by their call "Proletarier aller Länder, vereinigt euch!" ("Proletarians of all

---

Box 6.2 JULES VERNE AND PHILEAS FOGG'S WINNING OF HIS WAGER

One of the foremost fiction writers of the nineteenth century, Jules Verne (1828–1905) was fascinated with the technological innovations of his day and with how these were changing the geographical relationships between places. This is particularly evident in his 1872 novel *Around the World in Eighty Days*, in which he suggests that the shrinking of the globe by the spread of railways paradoxically made it both easier for a thief to escape his pursuers by being able to cross in shorter amounts of time greater distances from the scene of his crime and easier for his pursuers to find him, since the world was now that much smaller. Indeed, in the novel it is discussion of this crime which leads Verne's principal character, Phileas Fogg, to bet his friends that he can circumnavigate the world in eighty days. Leaving London at 8:45 p.m. on October 2, 1872, Fogg travels for 80 days and arrives back on what he thinks is December 21 at 8:50 p.m. – that is to say, 5 minutes late. However, in traveling eastwards rather than westwards, he had unwittingly gained a travel day and actually arrived back in London on December 20 at 8:50 p.m., after only 79 days. As Verne (1872/2004: 187) put it, in "going eastward, [Fogg] saw the sun pass the meridian eighty times, [whilst] his friends in London only saw it pass the meridian seventy-nine times."

countries, Unite!"). The idea that capitalism was, by nature, expansionary and would therefore develop at a global scale was also taken up by others drawing upon Marx's work, including Lenin (1926/1939), who saw imperialism as the highest stage of capitalism, and Luxemburg (1913/ 1951), who (incorrectly) argued that capitalism's geographically expansive nature meant that once all pre-capitalist corners of the planet had been incorporated into it, then capitalism would collapse in upon itself because it had reached its global limits and there were no more non-capitalist spaces into which it could spread (Box 6.3).

More contemporaneously, the past three decades or so have seen the global scale become a particular focus of attention within theorizing about "globalization" (Herod 2009). Thus, it has been frequently argued that national financial markets are "losing their separate identities as they merge into a single, overpowering marketplace" (Bryan and Farrell 1996: 4), and that this marketplace is increasingly shaped by the "operation of an international law of one price" (McKinsey & Company 2005: 7) in which the creation of a global market "based on a world in which borderlessness is no longer a dream . . . but a reality" represents "a process of global optimization" (Ohmae 2005: 18, 122). Such authors argue that globalization and the dominance of the global scale will bring about the creation of "one grand continent of opportunity whose exact contours still remain vague in places but that will amply reward brave exploration" (Ohmae 2005: 81) and result in a situation in which the globe is really only "intelligible as a *single place*" (Bartelson 2000: 187). As a consequence, according to Virilio (1995b), whereas previously "history has taken place within local times, local frames, regions and nations," globalization is "inaugurating a global time that prefigures a new form of tyranny." Thus, he has suggested, if in the past history was "so rich, it is because it was local, it was thanks to the existence of spatially bounded times which overrode something that up to now occurred only in astronomy: universal time." However, he has contended, increasingly "our history will happen in universal time, itself the outcome of instantaneity."

Virilio's assertion that the rise of a global scale of social life is eradicating others is certainly a common one within the globalization literature, and has been tied into ideas of simultaneity (Foucault 1986) and the annihilation of space by time. Arguably, the image which has been most associated with this narrative is that of the Global Village, first articulated by Canadian media critic Marshall McLuhan. As William Blake and

Box 6.3 LUXEMBURG, LEFEBVRE, AND THE SURVIVAL OF CAPITALISM

The Polish-German Marxist Rosa Luxemburg (1871–1919) argued in her 1913 book *The Accumulation of Capital* that capitalism requires non-capitalist markets to which it can export goods if it is to survive. As she put it (p. 416): "capital cannot accumulate without the aid of non-capitalist organisations, nor, on the other hand, can it tolerate their continued existence side by side with itself. Only the continuous and progressive disintegration of non-capitalist organisations makes accumulation of capital possible." Consequently, given that capitalism is an inherently geographically expansive mode of production, it would continue spreading spatially, she averred, until it had become global in its extent and there were no more non-capitalist spaces into which it could expand. However, Luxemburg argued, because it requires new spaces into which to expand, capitalism would come to a grinding halt once it had colonized the entire globe. Clearly, though, this has not happened. In an effort to explain why, French Marxist Henri Lefebvre (1973/1976) argued that capitalism has continued to survive through its ability to rework the spaces already within the capitalist system through a process of ongoing internal redivision. Thus, spaces that were previously developed are undermined through the geographical mobility of investment to underdeveloped places, which thereby develop at the expense of the former in what Smith (1984/1990) has called a "see-saw" motion of capital. As Lefebvre put it (1973/1976: 21): "capitalism has found itself able to attenuate (if not resolve) its internal contradictions for a century, and consequently, in the hundred years since the writing of *Capital*, it has succeeded in achieving 'growth.' We cannot calculate at what price, but we do know the means: *by occupying space, by producing a space.*

H.G. Wells had done, McLuhan also played with the idea of expressing the enormity of one scale – the global – by comparing it with something much smaller – the scale of a village. Hence, he contended, electronic media have "contract[ed] the world to a village or tribe" to the point "where everything happens to everyone at the same time [and] everyone knows about, and therefore participates in, everything that is happening the minute it happens" (Carpenter and McLuhan 1960: xi). In such a global village, electronic media have so compressed time and space, and the world and its peoples have become so interconnected, McLuhan suggested, that "'time' has ceased, 'space' has vanished" (McLuhan and Fiore 1967: 63). The result, he indicated (McLuhan 1964/2001: 67), has been a fundamental transformation in the nature – and spatial scale – of human understanding: "If the work of the city is the remaking or translating of man into a more suitable form than his nomadic ancestors achieved, then might not our current translation of our entire lives into the spiritual form of information seem to make of the entire globe, and of the human family, a single consciousness?" This is perhaps the ultimate expression of how the creation of a global scale of social and economic life is seen to augur a panoptical sense of time and space in which global history and geography are consumed, God-like, "at a glance – in a single spectacle from a point of privileged invisibility" (McClintock 1995: 37).

In this section, I have explored how material developments such as the spread of trade routes, colonialism, and technological improvements have all facilitated the creation of a global scale of economic and political organization. I have also highlighted some of the ways in which the global has been imagined in philosophy, theology, and popular culture. However, it is important to recognize that the creation of a global scale has not been a unidirectional process. For instance, the division of the planet into blocs associated with either West or East during the Cold War did much to undermine the greater integration of a global scale of social life. Likewise, much recent philosophical debate, especially that associated with postmodernism, with its distaste for metanarrative and its celebration of the local, has sought to destabilize the global as a scale of reference and ideational organization – that is to say, to destabilize how ideas of the global discipline our thinking. Nevertheless, the global remains important, both as a scale at which people seek to act and as a scale which holds significant sway over how they conceptualize the world and their place within it.

THE GLOBAL/LOCAL BINARY

Within both academic and non-academic analytical writing, the global has frequently been conceptualized as part of a binary. Freeman (2001: 1008), for instance, has claimed that several binaries have often been aligned when conceptualizing the global scale – global/local, masculine/feminine, production/consumption, formal/informal sectors of the economy. Specifically, she contends (p. 1012) that the local has commonly been portrayed as "contained within, and thus defined fundamentally by, the global" (à la, perhaps, a concentric circle representation of scalar hierarchy) and that the fusion of gender (specifically femininity) with the local has left "the macropicture of globalization bereft of gender as a constitutive force." The result has been a situation in which, although gender is not seen to be a constitutive element of globalization processes, the global, though formally ungendered, has in fact been taken as masculine because of the long tradition in Western thinking of associating maleness with the universal and the ungendered (as when the word "Man," though clearly gendered, is taken as the universal descriptor for all humans, such that it is made to appear that only women have gender). However, if taking the global as the (male) universal represents an implicit gendering, one very explicit gendering is that by Nagahara (2000: 935) who, drawing upon Marx's (1894/1981: 969) phraseology, has maintained that the process of globalization is one in which *Monsieur le Capital* ("Mr. Capital") and *Madame la Terre* ("Mrs. Landed Property") come together to facilitate accumulation. With *Monsieur le Capital* understood to be planetarily mobile whilst *Madame la Terre* is viewed as fixed place, this representation presents a dualism which plays on long-standing images of masculine as active and roving and feminine as passive and still. What is perhaps equally significant in all this is that such narratives not only portray the local and the global in quite gendered terms but that they are also decidedly heteronormative in their construction – the local and global as feminine/masculine are seen as opposites that "naturally" go together. Interestingly, Robertson and Khondker (1998: 37) claim that feminists themselves have often promulgated this dualism, suggesting that "there has been a strong tendency in some feminist circles to privilege the local and in fact to regard the discourse(s) of globalization as a masculine preoccupation," although Eschle (2004) disputes this.

As evidenced above, an important way in which the global has been theorized is relationally in terms of its contrast to the local, which is taken to be its Other (Box 6.4). What is noteworthy in this, however, is that although the local and the global have often been paired as two sides of a dualism, unlike the global, the local generally has not been theorized in particularly specific spatial terms, except to be seen as Other to the global. This may be because there is a geological limit to the global – the Earth's physicality extends only so far – whereas the local does not have such an obverse geographical limit. Nevertheless, whatever the reason, it is the case that the global has frequently been portrayed as having a geographically fixed upper spatial limit (the surface of the entire planet) whereas the local's lower spatial limit is seen as more geographically fluid. Significantly, within this dualism, both global and local have generally been associated with not just a masculine/feminine binary, but also binaries of abstract/concrete and powerful/weak, with the global associated with the first element in each of these two. In this regard, Gibson-Graham (2002) has identified at least six ways in which the relationship between the local and the global is regularly viewed.

1   The global and the local are seen not as things in and of themselves but, instead, as interpretive frames for analyzing situations. For instance, when considering processes of economic restructuring, what is seen from a "global perspective" (perhaps a worldwide economic slowdown) may appear different when viewed from a "local perspective" (some places may actually be experiencing economic expansion during such a global economic slowdown).

2   The global and the local each derive meaning from what they are not. Hence, much as our conception of what a slave is only makes sense if we can contrast that with what constitutes a free person, the global and the local only make sense when contrasted with each other. Thus, drawing on Dirlik (1999: 4), Gibson-Graham suggests that in such a representation the global is "something more than the national or regional . . . anything other than the local." In turn, the local is seen as the opposite of the global. This view, however, represents an important semantic shift, because whereas once the local "derived its meaning from its contradiction to the national," now anything other than the global is often seen to be "local" – for instance, nations and even entire extra-national regions (such as the European Union) are commonly referred to as "local actors" within the process of globalization.

Box 6.4 "OTHERING"

The notion of "Othering" has played an important part in contemporary critical theory, especially that related to feminism and postcolonialism, although its origins are much older. Hence, Hegel used the concept of "the Other" to describe a situation in which a consciousness sees itself in another and that this viewing plays a role in its developing self-consciousness. Such ideas have been reworked by many more contemporary writers. Thus de Beauvoir (1949/1953) argued that Woman has historically been treated as Other to Man in Western culture, with Man holding power in this binary – the Self/Subject is generally seen as the active and knowing (male) subject of knowledge whilst its (female) Other is passive and the known. In this relationship, she contended, the feminine Other is not an equal counterpart but, instead, is everything the Self/Subject rejects – it is passive rather than active, ephemeral rather than permanent, weak rather than strong. For his part, the psychoanalyst Jacques Lacan (1977) suggested that, at birth, every human is not yet a Subject because they are unaware of any other human against which to distinguish themself. However, he averred, gradually the Ego begins to gain self-awareness as it encounters Others, initially through a child seeing its own reflection in a mirror – which externalizes the child's body, driving a wedge between it and the Ego – and then through encounters with other individuals.

Arguably, in Geography it is the work of literary theorist Edward Said (1978) that has been most used to explore the concept of Othering. In particular, Said argued that there had been a long tradition in the West of using the East/Orient to help the West understand itself through the construction of myriad binaries. For example, the West has tended to see itself as progressive, democratic, and rational, but has done so through contrasting itself with what it has taken to be a conservative/backward, undemocratic/absolutist, and irrational Orient. Within this dualism, the West has bestowed upon itself what it sees as the more positive elements of these binaries, associating the less positive ones with its Other, the Orient. Although Said's interpretation has been criticized, it has nevertheless been important in spreading the concept of Othering.

3   Whereas many have viewed spatial scales in terms of a hierarchy of fixed and separate local and global arenas, in a networked conception of scale both the local and the global are not discrete realms but, rather, "offer points of view on networks that are by nature neither local or global, but are more or less long and more or less connected" (Latour 1993: 122). Thus, much as it would be impossible to describe a spider's web in hierarchical spatial fashion – where does one part end and another begin? – in Latour's view it is impossible to distinguish where the local ends and the global (or other scales) begins. Instead, the global and local are simply different "takes" on the same universe of networks, connections, abstractness, and concreteness. They are not so much opposite ends of a scalar spectrum but are a terminology for contrasting shorter and less-connected networks with longer and more-connected ones.

4   The global is local. In this perspective the global does not really exist, and if you scratch anything "global" you will find locality – multinational firms, for instance, are considered "multilocal" rather than "global."

5   The local is global, and place is a "particular moment" in spatialized networks of social relations. Hence, much as the tips of an octopus's legs might touch particular locations on the seafloor whilst its body floats above it, in such an approach the local is where global forces "touch down." In turn, the local is not a place but an entry point to the world of global flows encircling the planet.

6   The global and local are not locations but processes. Put another way, globalization and localization produce all spaces as hybrids, as "glocal" sites of both differentiation and integration (Dirlik 1999: 20). Thus, the local and the global are not fixed entities but are always in the process of being remade. Local initiatives can be broadcast to the world and adopted in multiple places, whilst global processes always involve localization (for instance, in the process of globalizing itself McDonald's tailors its products to particular local tastes).

In reviewing these different ways of articulating the relationship between the local and the global, Gibson-Graham (2002) suggests that the history of this binary has been one in which the global's power has usually been assumed to be greater than the local's. Part of this results from the widely held view in Western thought that greater size and extensiveness imply domination and superior power, such that the local is often represented as "small and relatively powerless, defined and confined by the global" (p. 27). In such a representation "the global is a force, the local is its field of play . . . the global is penetrating, the local penetrated and transformed." Thus, the global is conceived of as "synonymous with abstract space, the frictionless movement of money and commodities, the expansiveness and inventiveness of capitalism and the market. But its Other, localism, is coded as place, community, defensiveness, bounded identity, in situ labor, noncapitalism, the traditional." Such a standpoint has been articulated by, amongst others, Lefebvre (1978/2009: 238), who suggested that the nation-state is caught in "the collision between two practices and two conceptions of space, one logical (global [globale], rational, homogenous), the other local (based on private interests and particular goals)" – a configuration in which he locates the national scale between that of the logical global and the, by implication, illogical local. Likewise, Chang and Ling (2000: 27) use the stark term "techno muscular capitalism" to describe an "integrated world of global finance, production, trade and telecommunications" in which "all those norms and practices usually associated with Western capitalist masculinity – 'deregulation,' 'privatization,' 'strategic alliances,' 'core regions,' 'deadlands' [– and which are presented] as global or universal" are valorized, in opposition to local productive forces. What emerges from this, Gibson-Graham (p. 33) argues, "is an overriding sense that [power] is either already distributed and possessed or able to be mobilized more successfully by 'the global'" than by the local.

Paradoxically, both the political right and the political left have routinely vaunted the global's power. Thus neoliberals regularly use the rhetoric of globally organized capital's command of space to undermine local opposition to their agenda (as in the myriad claims that local union work rules "must" be conceded "due to the imperatives of global capitalism"), whereas many on the political left appear to have given up any hope of challenging capitalism globally and prefer instead to "think globally" but to "act locally." However, Gibson-Graham argues that such a discursive reification of the global restricts possibilities for progressive political action because it suggests that capital (viewed as being global in its operation) can always outmaneuver its opponents (who, as Other to capital, are usually viewed as non-global or local). Instead, Gibson-Graham has sought to deconstruct the local–global binary to explore how the local can perhaps be thought of as enabling political struggle and as a potentially powerful basis for challenging (global) capital and how the global, in turn, may be understood and represented in ways that do not assume its omnipotence. Hence, she suggests, organizing locally can indeed be effective in particular circumstances (for an example, see Herod 2000) and that we should not think of the global as a scale at which only capital can operate effectively (see Herod 1997c). Equally, Barber (2001) has suggested that the growing power of the global actually facilitates the growing power of the local, as the global spread of Western culture encourages a heightened defense of local cultures in response – a phenomenon which he calls "Jihad vs. McWorld" and which Rosenau (1999: 293) has called "fragmegration," a term which he argues captures "the inextricably close and causal links between globalization and localization, highlighting the possibility that each and every increment of the former gives rise to an increment of the latter, and vice versa." Finally, understanding how the categories global and local are constructed discursively can have signal import for the politics of organizing – viewing a transnational corporation as "multilocational" rather than as "global," for instance, can significantly shape workers' and others' political calculations.

Significantly, much of the work in this vein has taken the existence of the global and the local somewhat for granted – that is to say, the questions asked are less about how they came to be and more about what is their relationship to one another. Some have taken a slightly different tack. Smith (1984/1990: 139), for instance, has focused upon how the global scale has been actively built and transformed as the forces of capitalist uneven development have developed. Hence, he has argued, the transition from a world that was beginning to be more fully integrated by trade beginning in the sixteenth century to one which has been increasingly integrated by the transnationalization of production chains has meant that whereas industrial capital "inherit[ed] the global scale in the form of the world market . . . based on exchange," it has transformed it "into a world economy based on production and the universality of wage labour." Massey (2004) likewise has insisted that the local and the global are equally tangible and socially produced, a position which challenges the asymmetry in many binaries – the global as abstract, the local as concrete, for instance. Thus, she maintains that

> if space is a product of practices, trajectories, interrelations, if we make space through interactions at all levels, from the (so-called) local to the (so-called) global, then those spatial identities such as

places, regions, nations and the local and the global, must be forged in this relational way too, as internally complex, essentially unbound-able in any absolute sense, and inevitably historically changing.

(p. 5)

What this means is that

if we take seriously the locally grounded nature even of the global, and take seriously indeed that oft-repeated mantra that the local and the global are mutually constituted, then ... local places are not simply always the victims of the global; nor are they always politically defensible redoubts against the global. [Rather,] places are also the moments through which the global is constituted, invented, coordinated, produced. They are 'agents' in globalisation.

(p. 11)

The result is that "the inevitably local production of the global means ... there is potentially some purchase through 'local' politics on wider global mechanisms[, n]ot merely defending the local against the global, but seeking to alter the very mechanisms of the global itself" (p. 11). At the same time, however, "the identities of places are ... the product of relations which spread way beyond them" (p. 11). However, in viewing the local and the global as mutually constitutive of one another, Massey (p. 13) maintains that "the very inequality inherent within capitalist globalisation [means] that the local relation to the global will also vary, and in consequence so will the coordinates of any potential local politics of challenging that globalisation" – in other words, there is a historical and geographical unevenness in the relationship between the local and the global.

There are three final issues to address when considering the relation-ship between the global and the local. The first is that it is important to recognize that the global itself as a category has been viewed in different ways by various scholars. For his part, Cosgrove (1994) has detailed how the global has been presented in two quite dissimilar fashions: in terms of the rhetoric of a "one-worldism" and that of a "whole-Earthism." Although both terms have been used to describe a sense of unified globalism, they can be read in two distinct ways. Hence, "one-worldism" suggests a much greater degree of integration and global singularity than does "whole-Earthism," which can be taken to imply that the Earth is made up of myriad separate, though conjoined, parts. Meanwhile, French theorist Lefebvre distinguished between the *global*, which he saw as referring to a level of analysis, and the *worldwide*, which he saw as a material geographical scale (Brenner and Elden 2009: 22). Nevertheless, he viewed the global as part of a "hierarchical stratified morphology" wherein "the local, the regional, the national and the global imply one another" (quoted in Brenner 1997: 135) but in which "contemporary capitalism can only be understood adequately on a global scale, in terms of the encompassing space of the world system, the final spatial frontier for capital" (p. 143). Whereas some have suggested that the global erases other scales of existence, Lefebvre was adamant that subglobal scales are fundamental elements of globalization: "*the worldwide* [le mondial]," he professed, "*does not abolish the local.*" The result is that humanity is "confronted not by one social space but by many . . . by an unlimited multiplicity or uncountable set of social spaces," such that "the global scale must be conceived as a 'hypercomplex,' 'polyscopic,' and 'contradictory' amalgamation of multiple forms of sociospatial organization, not as a reified, territorialized essence" (p. 144).

Second, in an attempt, perhaps, to bridge the local/global divide, a number of writers – within Geography, most notably Swyngedouw (1997b), but also others – have pushed the concept of a glocal scale and glocalization to represent the co-presence of both the local and the global and to suggest that these two scales mutually constitute one another. The glocal, suggests Gabardi (2000: 33), emerges out of the "development of diverse, overlapping fields of global-local linkages [which have] created a condition of pan-locality." However, such a representation presupposes that the global and the local pre-exist the glocal – it is the already extant global and the local which must come together to form the glocal. Hence, this *portmanteau* word contains within it a particular notion of both the local and the global as separate entities that are then brought together – it represents, according to Latham (2002: 138–39), "a case of mixing up scales." In contrast, Latham suggests that it is better to think of the relationship between the local and the global in networked terms. Hence, he avers, "a topologically oriented approach offers . . . an avenue through which to break down apparent divisions between scales into more analytically useful accounts." In particular, he declares, "[d]rawing on the geometric language of topology allows us to disassemble the global – to see it in its particularity, to see that it is provisional and not some monolith over which we can have no influence."

Third, whereas both neoliberal and leftist writers have often presented the global as a scale at which economic and political life inevitably will increasingly be organized – Ohmae (2005: 18), for instance, has argued that the global's power "is going to grow stronger rather than weaker," whilst Bauman (1998: 1) has suggested that the emergence of a global scale of capitalist organization "is the intractable fate of the world, an irreversible process" – in fact, recent events suggest that the global's power, at least as measured by its alleged undermining of other scales' influence on social life, is not so hegemonic. Indeed, the recent resurgence of the nation-state in the context of the Global Financial Crisis that began in 2007 speaks to this point precisely. Hence, the various government bailouts and take-overs of private institutions and moves by national governments to impose new regulations on globally mobile financial capital suggest that the rhetoric of the global scale's supremacy is overblown and that we may be entering a period of mounting national-level regulation of markets, in much the same way that the growing integration of the global economy in the late nineteenth/early twentieth century was reversed by the protectionism of the Great Depression (Herod 2009).

SCALING GLOBAL CITIES

In recent years, "global cities" have come to be seen as archetypical places where the local and the global – and even the regional, in the case of so-called "global city-regions" – come together. Typically, global cities are those cities which are seen to play significant roles in shaping and articulating the global economy, as if they were its command centers. They are understood to "act as functional nodes in global flows and networks of production, exchange, and consumption" (Ward and Jonas 2004: 2122). Possibly the first writer to comment in such terms was Geddes in his 1915 Cities in Evolution, when he explored how cities such as Paris, Berlin, and New York were growing and becoming more important as centers of worldwide commerce. However, one of the earliest explicitly to use the word "Global" in this regard was Friedmann (1986), who suggested that, at the time, New York, London, and Tokyo were in the first ranking of the planet's cities (they were "global financial articulations"), cities such as Los Angeles, Miami, and Singapore were second rank "multinational articulations," and cities such as Paris, Seoul, and São Paolo were third-ranked "important national articulations." These three levels, he argued, collectively formed a planetary urban network. Below these, at the other end of the scalar continuum, are myriad smaller places which are not considered "global." One of the significant issues to emerge out of Friedman's piece is that it highlights an important discursive change which appears to have taken place in recent years with regard to such cities, one associated with the broader purchase that the idea of the global has secured over the contemporary imagination – whereas today cities such as London and New York are frequently referred to as "global cities" (e.g., see Sassen 1991), previously they had generally been termed "world cities" (e.g., Hall 1966; Friedmann and Wolff 1982) or "international cities" (Barlow and Slack 1985), although this discursive switch is not total (Taylor's 2004 book uses the word "World" in the title). In some ways, this nomenclatural shift seems to have mirrored that in other arenas, as with the growing use of the word "Global" in the names of planetary organizations relative to the previously favored "International" or "World" (Herod 2009: 48–49).

For his part, Friedmann saw such cities as emerging out of the new international division of labor that developed in the 1970s. Hence, some cities were becoming headquarter cities for transnational corporations and so played the role of "articulator" cities, the portals through which a national or regional economy is linked to the larger global one – a representation that tended to see such scales as discrete entities. For Friedmann, urbanization was linked to broader global economic forces and global cities lay "at the junction between the world economy and the territorial nation state" (Friedmann and Wolff 1982: 312) – a significant reworking of the more traditional view amongst many that the urban scale is below, or contained within, the global and national scales. Friedmann also suggested, in global–local terms, that not only how a city's local economy is structured shapes its role within the broader global economy but also that its role in the broader global economy shapes how local labor markets function. By the 1990s, though, a narrow view of cities' characteristics as shaped by their place in the international division of labor had given way to a broader engagement with the literature on globalization. Hence, Sassen (1991: 3–4) argued that what distinguished truly global cities such as London, New York, and Tokyo from others was the fact that they are the command points for the global economy, that they have become key locations for financial and business services (which have replaced manufacturing as the leading economic sectors), that they are the leading sites for the production of innovations in these sectors, and that they are also thus becoming the leading marketplaces for such innovations. Nevertheless, although the appellation World City has been challenged by that of Global City, and although the specific cities involved change both historically (depending upon their economic or cultural development) and according to whomever is writing, the general argument remained the same, namely that there is a hierarchy of cities – based on size, economic and political function, and cultural characteristics – within the panoply of urban areas across the Earth's surface, some of which are more global than others.

In considering recent developments, Taylor (1997: 324) proclaimed that global cities "are the organising nodes of world wide networks." In this regard, they are not necessarily new – arguably, one could claim that the Rome of two millennia ago served such a function, as did Italian cities such as Genoa during the sixteenth century, Amsterdam in the seventeenth, and London in the eighteenth and nineteenth. However, whereas prior to the eighteenth century there was generally a single city at the top of the global hierarchy, and during the era of formal imperialism in the nineteenth and early twentieth centuries imperial metropoles such as London, Paris, and Berlin "each had their own separate imperial urban hierarchies," with inter-urban networks largely contained within various nation-state territories, for Taylor what is new about developments during the twentieth century is that "economic globalization has created the conditions for the development of a single worldwide urban hierarchy." It is this "world hierarchy hypothesis, seen as the geographical dimension of economic globalization, which is unique to contemporary world cities" and their study (p. 329). For Taylor, the key difference is that the myriad national urban hierarchies of yesteryear have now become intertwined in a unitary global network (Box 6.5).

What the world/global city thesis reflects, then, is the opinion that the planet's cities are increasingly being ordered within an integrated urban system that is only viewable and comprehensible from a single, privileged, global point of view. Moreover, it intimates that such global cities have broken free of the other scales within which they were previously contained. Hence, according to Taylor (p. 329), a "fully developed world city hierarchy implies an integrated global economic space

---

Box 6.5 THE GLOBALIZATION AND WORLD CITIES RESEARCH NETWORK (GAWC)

The study of Global/World cities has been formalized through the creation of the Globalization and World Cities Research Network (GaWC). The GaWC is based out of the Geography Department at Loughborough University (UK), but has strong links with the Metropolitan Institute at Virginia Tech (USA), the University of Ghent (Belgium), the Chinese Academy of Sciences, and several other institutions around the world through its affiliated academic members. The network produces research findings on global cities, as well as pedagogical sources for teaching about them. These are available at www.lboro.ac.uk/gawc.

---

that has overcome the obstacles represented by the political spaces of [nation-]states. Such a change . . . is epochal." At the same time, alongside "globalizing tendencies that promote a world city hierarchy there are continuing 'nationalising' tendencies that are countering [its] development." The result is that "there is currently severe tension between continuing nation-state reproduction and economic globalization[,] with both impinging upon each other: as well as globalization constraining states, states continue to be obstacles on the road to a uniform world economic space." Consequently, the "development of a world city hierarchy out of economic globalization is . . . neither a simple evolution nor . . . an inevitable outcome of contemporary economic and political changes in the modern world-system" (p. 330). Finally, Taylor (p. 323) contended, although a "world city hierarchy is a common assumption underlying this field of research," it seems to have "drawn the short straw when it comes to rigorous research." Hence, he maintained, for much global cities research

the idea of a hierarchy is left floating in the background, something vaguely obvious, presumably on the grounds that cities are inherently hierarchical in their relations. Historically, there is much to be said for this presumption, but in a situation where claims are made for categorically new urban processes, the idea of a hierarchy needs fresh empirical grounding and theoretical interpretation.

Within this global cities literature there are several scalar issues worth contemplating. The first is the suggestion that there has been an important realignment of the scale of the urban in relation to the national and the global. Hence, Abrahamson (2004: 5) has argued that economic developments in recent years have led to a "transition from industrial social organization in which the primary integrative role of cities was intra-national to a post-industrial order in which cities played a more important international role . . . Global cities were one of the many new forms to emerge out of this complex set of economic and cultural changes." Likewise, Sassen (1991: 3) has maintained that recent developments in the world economy have "created a new strategic role for major cities" vis-à-vis nation-states, whilst Isin (2001: 351) posits that the fact that "cities and regions, or more precisely, global city-regions, are the fundamental spaces of [the global economy] further erodes the credibility of modernization theories that would have us believe in national trajectories" of economic development. Meanwhile, Friedmann (1995) has contended that cities and inter-urban networks are supplanting nation-states as the basic territorial units of capitalist accumulation within what had previously been seen as an international economy organized into nation-states (Radice 1984) – they are, as it were, breaking out of the nation-state containers within which they had previously been enclosed. Such an apparent merging of the urban and the regional scale *via* the emergence of global city-regions led Haggett (1972/1975: 420), nearly four decades ago now, to suggest that the "arguments for adopting the city region as the basic spatial unit are persuasive," given that more and more humans are living in cities, that humans' "organization of the globe is increasingly city-centered," and that "[c]ities form easily identifiable and mappable regional units."

However, in making such claims it is important to consider how global cities have been presented scalarly and conceptualized as analytical units. Ward and Jonas (2004: 2124), for instance, have suggested that there has been "a tendency to reify city-regions as rational economic actors in a space economy comprised principally of trade and capital flows between functionally connected global-regions." For them, this has been the result "of privileging an exchange-relations reading of capitalism [which] effectively reduces cities, with all their institutional, political, cultural, and social complexity, to nodal points in a spatial cost surface that extends on a global scale." The result is that "theorists working from this position [have] tend[ed] to focus on the macroeconomic function and emergent properties of the city-region as an already-constituted functional collective having little or no formal connectivity to wider national geographies" – by talking of "cities as regions rather than cities in nations" (p. 2125) they have denied the gestalt of scale to see global cities as somehow analytically detached from national and other spaces.

For his part, Brenner (1998b: 2) has explored how descriptions of global cities' emergence have privileged the global as an analytical scale and thereby "deflected attention away from the crucial role of the state scale in the currently unfolding transformation of world capitalism." By primarily conceptualizing globalization as a process whereby a somewhat passive nation-state is undermined, Brenner averred, researchers have focused "on the global scale, the urban scale and their changing interconnections while neglecting the role of [nation-]state-level processes in the current round of capitalist restructuring" (p. 8). Such an approach sees these two scales – the global and the national – as mutually exclusive: "what one gains, the other loses." However, whereas global cities research has "generally presupposed a 'zero-sum' conception of spatial scale which leads to an emphasis on the declining power of the territorial state in an age of intensified globalization [wherein] the state scale is said to contract as the global scale expands" (p. 3), Brenner instead argues that nation-state territorial power is not being eroded but is, rather, being "rearticulated and reterritorialized in relation to both sub- and suprastate scales." The resultant "re-scaled configuration of state territorial organization," he maintains, "can be provisionally labeled a 'glocal' state."

Consequently, for Brenner global cities are not superseding the national scale but are "sites of reterritorialization for post-Fordist forms of global industrialization" and, as such, "coordinates of state territorial organization" – they are "local-regional levels of governance situated within larger, reterritorialized matrices of 'glocalized' state institutions" that play a central role in capital accumulation within the global economy. What this formulation means is that global cities "are not to be conceived as uniquely globalized urban nodes within unchanged national systems of cities and state power, but rather as sites of both socioeconomic and institutional restructuring in and through which a broader, multi-scalar transformation in the geography of capitalism is unfolding" (p. 12). The result is that they cannot be understood "without an examination of the matrices of state territorial organization within and through which [their formation] occurs[, for the] globalization of urbanization and the glocalization of state territorial power are two deeply intertwined moments of a single process of global restructuring through which the scales of capitalist sociospatial organization have been reconfigured since the early 1970s" (p. 27). Consequently, he claims, "globalization must be understood as a re-scaling of global social space, not as a subjection of localities to the deterritorializing, placeless dynamics of the 'space of flows'."

The second issue that emerges from this discussion is that the language of global city research has very much been that of networks, with global cities seen as nodes within an integrating global economy. Although several authors have taken up this idea, arguably it is Castells who is most closely associated with it. For Castells (1996: 386), contemporary developments – particularly the emergence of "the network society" – have transformed cities, such that the "global city is not a place, but a process . . . by which centers of production and consumption of advanced services, and their ancillary local societies, are connected in a global network, while simultaneously downplaying the linkages with their hinterlands." Castells (1999/2002: 372) suggests that much thinking on urban geography – especially that which has conceptualized "global cities" as somehow floating above other scales of social life, like the nation-state, regions, and "regular" cities – has been a captive of "a nineteenth-century, hierarchical conception of . . . society and space." By way of contrast, he declared, what characterizes contemporary society is "its structure in networks and nodes, not in hierarchies of centrality and periphery." What this means, for Castells, is that "no city is entirely global" – as he puts it, Queens in New York and Hampstead in London are "very local." At the same time, local places such as Wall Street and the City of London "interact in a global network," as do many smaller metropolitan areas to varying degrees. In fact, Castells argued, "hundreds and thousands of localities are connected in global networks of information-processing and decision-making," such that "[a]ll large metropolitan areas in the developed world, and all of the largest in the developing world, are thus global to some extent, with their relative nodal weight in the network varying depending upon time and issues." However, because "most people in these same cities live local lives . . . there are not [multiple] global cities . . . but one global city [which] is not New York or London [but] a transterritorial city, a space built by the linkage of many different spaces in one network of quasi-simultaneous interaction that brings together processes, people, buildings, and bits and pieces of local areas, in a global space of interaction." The global city is not so much a city but "a new spatial form, the space of flows, characterizing the Information Age."

The third issue (which relates to, though is somewhat different from, the second) is the emergence of a discourse concerning global cities which replaces the language of scale with that of "folds." Arguably, the clearest exposition of this shift is in R.G. Smith's (2003) writing, wherein he rejects scale as too static a concept and instead engages with post-structuralist authors such as Gilles Deleuze, Félix Guattari, and Bruno Latour. Drawing on the imagery of origami, Deleuze in particular outlined how space could be thought of in terms of folds and refolds, rather than Euclidean geometries of spatial containment. As Smith (p. 565) explains it, "[w]hat interests Deleuze are (un)folds, the infinite labyrinth of fold to fold that produces the world's topology as one of process that overwhelms the fictions of boundaries, limits, fixity, permanence, embedment." Similarly, Serres (1991/1995: 60) has used the image of the handkerchief to explore space in terms of folds rather than scales, indicating that if one were to "take a handkerchief and spread it out . . . to iron it, [one could] see in it certain fixed distances and proximities . . . [whereas taking] the same handkerchief and crumpl[ing] it" means that two points which were distant when the handkerchief was flat "suddenly are close, even superimposed." Moreover, if the handkerchief is torn in certain places, "two points that were close can become very distant." Calling this "science of nearness and rifts" "topology," and distinguishing it from what he called "metrical geometry" ("the science of stable and well-defined distances"), Serres likewise presented a view in which spaces emerge and are constantly reconfigured as socio-material relations "arranged into orders and hierarchies" which are fluid, rather than fixed in any Cartesian sense.

This is important, Smith avers, because writers such as Sassen have tended to see geographical scales such as the local, the national, and the global as distinct, if overlapping, entities and therefore have thought about global cities in terms of discrete scales, boundaries and territories – she has "a scalar view of globalization so that when scales (and so boundaries) meet they 'overlap' producing 'new zones' that are in need of research" (p. 570). By way of contrast, and drawing upon Doel's (1999: 18) comment that "the world can be (un)folded in countless ways, with innumerable folds over folds, and folds within folds," such that "no folds . . . become redundant, nor [can any] of them seize power as a master-fold [so as to] be distinguished in terms of the essential and the inessential, the necessary and the contingent, or the structural and the ornamental," Smith (p. 570) contends that his own ontology of globalization "fluidifies such solidified thinking revolving around such motifs as fluidity and flow, movement and mobility, folds and networks." The result is "a rejection of scales and boundaries altogether as globalization and world cities are too intermingled through scattered lines of humans and non-humans to be delimited in any meaningful sense."

Finally, viewing global cities and other urban places as networked together has two interconnected consequences for thinking about the urban as an object of study, especially as manifested in the form of a global city. First, as Amin and Thrift (2002: 1) ask,

> [if] the city is everywhere and in everything[, if] the urbanized world now is a chain of metropolitan areas connected by places/corridors of communication (airports and airways, stations and railways, parking lots and motorways, teleports and information highways)[,] then what is not the urban? Is it the town, the village, the countryside? Maybe, but only to a limited degree[, for the] footprints of the city are all over these places, in the form of city commuters, tourists, teleworking, the media, and the urbanization of lifestyles.

The result is that the "traditional divide between the city and the countryside" is seen to have been "perforated," a view which challenges long-standing notions of what it means to talk about the urban scale going back to analysts such as Wirth (1938) and Marx and Engels (1845/1970) (see Chapter 3). If cities "no longer have defined edges," Amin and Thrift (2005) ask, if they are "the parentheses in the flows of the world . . . and if much of their rationale derives from their connections with other places," then analytically "how can we find an object to grasp?" In other words, what exactly is the object of analysis when considering a globalized urban within a space of flows?

Second, given that global cities are networked together and their development is predicated, at least in part, on their connectivity with other cities, such that there can be, according to Taylor (2004: 8) (following Castells), "no such entity as 'the city', meaning a single city," it is necessary to think of all cities in relational terms. Such an approach, however, challenges researchers who have "become accustomed to seeing cities as urban places, complete with boundaries, within their respective countries" and, Taylor avers, makes it virtually impossible to think of the urban scale in discrete, areal terms. At the same time, such a repudiation of the urban as a discrete, areal scale – a view which Taylor believes emerged as the nation-state strengthened as a territorially discrete entity – does not mean that other scales might not continue to be discrete and areal for, he suggests, cities and nation-states represent two different types of space: "cities in a network are a space of flows, whereas nation-states forming a territorial mosaic (the world political map) are a space of places" (p. 27). For Taylor, the issue is not whether networks are or are not replacing in toto areal views of scale. Rather, it is that these alternative forms of space express different social processes – "the nationalization of humanity over the past two hundred years [and] the global challenge to this prioritization of places over flows today." Consequently, "contemporary globalization can be viewed geographically as a tension between two 'world spaces': network and mosaic" (p. 27–28).

In considering how global cities have been thought of as networked, Leitner and Sheppard (2002: 497) have commented that, although there are exceptions, within post-structuralist analysis networks are often seen as: self-organizing (they "evolve a relational organizational structure that is bottom-up, rather than externally imposed[, they] are path-dependent and evolutionary, and unpredictable in the medium run because network dynamics can be dramatically affected by small changes in external conditions"); as collaborative; as non-hierarchical; as flexible (continually subject to change and periodical restructuring as participants come and go, and thus possessed of fuzzy boundaries); and as having a spatiality that is topological ("networks evolve by creating linkages between participants who were not previously connected, thereby constructing mutuality between . . . actors, or places, that previously seemed distant from one another"). Networks, post-structuralists such as Latour (1993: 118) frequently posit, thus "are nets thrown over spaces, and they retain only a few scattered elements of those spaces. They are connected lines, not surfaces." In reality, however, at least based upon their research within the European Union, wherein myriad networks have been created between cities as a way to transcend national boundaries and link specific urban locales supranationally, Leitner et al. (2002: 495) have demonstrated that actual interurban networks depart significantly from the utopian discourses with which they are frequently described. Thus, because networks are "[e]mbedded within pre-existing processes of uneven development and hierarchical state structures," they exhibit "internal power hierarchies[,] are created, regulated, and evaluated by state institutions, and often exclude institutions and members of civil society." Equally, although "they create new political spaces for cities to challenge existing state structures and relations," these "are of unequal potential benefit to participating cities."

For his part, McCann (2002: 63–64) has contended that there has been "an increasingly clear and troubling bifurcation in urban studies literatures between, on the one hand, research on the world's largest cities (and attendant writing intended to show how other cities display signs of 'globalness' . . .) and, on the other hand, research on what are implicitly deemed to be other 'ordinary' or 'local' cities." In such a discourse, cities such as London or New York are used as templates to measure other cities' globalness and to understand urban processes elsewhere. As a consequence, "places that do not vividly express similar processes to those seen in the 'global cities' are deemed to be less affected by globalization and are less likely to be seriously studied." The result, he declares, is a "problematic deficiency in our understanding of the social processes that shape these supposedly 'nonglobal' or 'less-global' places and the lives of those who live in them." Through such approaches "the urban" as an object of study is increasingly being remade as less a relatively self-evident scalar entity and more as a node within the interscalar networks and supraurban scalar hierarchies which global cities are seen to inhabit. "The persuasive rhetoric of the global cities literatures," McCann pronounces (p. 67), must therefore "be seen as a reframing of the urban question in a manner through which certain cities are articulated at a new scale – archetypal globalized cities – while the rest are defined (by omission) as less important to our understanding of contemporary urbanism and global restructuring." The outcome is the representation of the "global as a separate sphere of activity from the local" (p. 75) – and, we might add, other scales too.

In contrast, McCann seeks to cultivate a "language for understanding place [that will] prove helpful to the development of an urban studies that does not privilege the global as a scale of analysis" (p. 64). He does so by returning to dialectics to suggest that the local and the global are not separate arenas and part of a dualism but, rather, that through the notions of "multiplexity" and "co-presence" all cities – "global" and "nonglobal" – can be understood as places in which diverse economic, political, and cultural networks are co-present and interact in complex and contradictory ways. Hence, every city is understood to be a unique bundle of the broader processes driving uneven development under capitalism, with the notions of multiplexity and co-presence providing a perspective on place "that emphasizes *both* the uniqueness *and* the interconnectedness of places" (p. 77). Such a perspective, McCann (2004: 328) contends, sees the global and local "as powerful but historically contingent crystallisations of fluid social processes that transcend the boundaries of categories such as 'global' and 'non-global'."

Likewise, M.P. Smith (2001: 50–51) has argued that the global cities literature, in which "the grand narrative of capitalist urbanization" presents global cities as having "a single, unitary author: transnational capital," has tended to reify the global city "as a fabricated by-product of the structural transformations of global capitalism in the late twentieth century." One way in which this happens is in how the "self-evident" global cities of New York, London, and Tokyo are frequently used as the benchmarks for how global are other cities across the planet. However, Smith contends, "global city assumptions about the systematic coherence of the urban hierarchy, the transterritorial economic convergence of global command and control functions, and the declining significance of the nation-state, are more difficult to maintain" than the literature portrays. Furthermore, in deconstructing Sassen's argument that local factors are facilitating the growth of global cities and that this growth is redefining the relationship between city and nation-state, Smith (p. 56) asserts that Sassen sees global cities as "globally and locally embedded but nationally disembedded." In fact, he suggests (p. 58), because much of the global cities literature "depends on the assumption that global economic restructuring precedes and determines urban spatial and sociocultural restructuring, inexorably transforming localities by disconnecting them from their ties to nation-states, national legal systems, local political cultures, and everyday place-making practices," the globalization and global cities narrative actually helps create the sense of local "powerlessness that it projects by contributing to the hegemony of prevailing globalization metaphors of capitalism's global reach, local penetration, and placeless logic."

CONCLUDING REMARKS

In this chapter, I have tried to do three things: to explore how the global has been historically constructed both materially and discursively; to illustrate how the global has frequently been seen as part of a binary, and with what effect; and to delve into how the rise of a literature on global cities ties into the globalization literature and what it means for considering the relationship between the global scale and others, especially the urban and the national. What is evident from such a discussion is that the moves toward building a more interconnected global scale of social life have been decidedly uneven, both geographically and historically. Hence, although capitalism has gradually become more globally organized, it has done so in fits and starts. Thus, for instance, the construction of the telegraph in the nineteenth century, so crucial to establishing a material connectivity across the globe but also to fostering a sense of globalism, only connected certain places together, and did so in a particular order – the US and the UK were connected before the UK and Australia or India were, for instance. Likewise, the emergence of various so-called global cities as alleged control points of the global economy has had a particular historical geography to it – not all places have been impacted simultaneously.

Such considerations of the global's origin and how it is conceived are important because the global has often been seen as the acme of scales, the scale which either encompasses all others or sits atop them and from which there is no hope of disengagement. Moreover, within the "globalization talk" of both neoliberals and many who oppose neoliberalism, the global has generally been represented as having become ever more powerful over the past three decades or so, at the expense of other scales. However, with growing international labor solidarity and other transnational social movement actions (such as those that have come together to oppose the World Trade Organization's trade talks), high oil prices (which have encouraged some offshoring manufacturers to repatriate work – so-called "backshoring"), and, more recently, significant material changes unfolding within the global economy stemming from ongoing financial and manufacturing crisis, long-standing representations of the global as the master scale of contemporary capitalism have increasingly been challenged. Likewise, more critical evaluations of the role of, for example, nation-states in shaping globalization show that, despite the early triumphalist neoliberal rhetoric in which the global was often heralded as eradicating the nation-state, global markets can usually only be captured through political control of various nation-states, who are the ones who must sign into law treaties guaranteeing "free markets" – in other words, domination of the global usually rests upon domination of some other scale. Hence, as Polanyi (1957: 139–41) reminds us, there is nothing natural about neoliberal globalization, for "free markets could never have come into being merely by allowing things to take their course ... [L]aissez-faire itself was enforced by the state ... Laissez-faire was planned." What is more, because globalization and constructing a global scale of economic organization are not pre-ordained but are chosen – they are "a deliberate choice, rather than an ineluctable destiny" – they do not necessarily render states impotent, for states' "potency lies in the choices they make" (Wolf 2001: 182–83).

The growing use of a language of networks to challenge an areal view in which the global is seen to surmount and/or enclose all others in a vertical or horizontal hierarchy has also impacted how the global is currently being (re)conceptualized. For example, whereas a narrative of enclosure or vertical hierarchy long dominated thoughts about the global's relationship to other scales, now a discourse of nearness and farness has become widespread, such that both discourses are now commonly used to describe the contemporary scene. Nevertheless, although in many ways the language of networks is quite different from that of scales as areal units, there are similarities in both sets of discourses. In particular, in both areal and networked discourses the global and the local have often been seen as Other to one another, with additional sets of dualisms superimposed upon these – the global as powerful, abstract, ubiquitous, and large, the local as weak, concrete, specific, and small, for instance. Such dualisms are problematic, for they fetishize scales – they suggest that particular scales by definition and nature exhibit certain characteristics. Part of the challenge of thinking about the global scale critically, therefore, is to avoid such simplistic dualisms.

Finally, in considering the global as a scale, it is important to contemplate critically the relationship between social actors' material practices and how these are described discursively. For instance, what is interpreted as a firm "going global" with regard to its investments will depend, to a certain extent, upon how the global is understood – does it mean having a plant in every large city of the world, merely one in every country or just one in every continent, or something else? Equally, determining which is a global city and whether the number of such cities is increasing "as a result of globalization" is dependent upon how a global city as a concept is defined. Although changing the language with which different scales are described does not change the materiality of such scales, it can dramatically affect how they are understood and how social actors behave in response to such understandings. Hence, as already mentioned, Gibson-Graham has remarked that something as simple as describing a transnational corporation as "multilocational" rather than as "global" can have significant political impacts – for workers pondering a challenge to such a company, facing what is perceived to be a "multilocational" firm no doubt seems much less challenging than facing one that is understood to be "global." Likewise, a firm might manipulate representations of its organizational structure, presenting itself as "global" when faced down by striking workers but as "local" when seeking tax breaks from munici- palities or when trying to convince consumers in a country different from that in which it originated that it is "not really" a foreign company – as Toyota, for instance, has sought to do in the US through emphasizing how many American workers it employs and how much it contributes to the US economy. The ability of social actors both to construct materially their scales of organization at different spatial resolutions and to represent their organization discursively in different ways – some of which may directly mirror their material geography of organization and some of which may be actually quite different from it – is a central element in political praxis.

7

CONCLUSION

In this book I have explored various ways in which geographic scales have been conceived – as areal "space envelopes," as networked, as material social products, as mental fictions, as merely logical divisions of the Earth's surface, and so forth – together with how they have been represented discursively. I have focused on five scales in particular – those of the body, the urban, the regional, the national, and the global – to understand how each of these has been conceptualized in and of itself and what this means for thinking about each as a geographical scale (for instance, how does the way in which we think of the human body as a body shape how we think about the body as a scale?). I certainly recognize that these are not the only scales upon which I could have focused, but I would also argue that they are probably the five which have been most central in debates within Human Geography and which, given space constraints, it probably makes sense to highlight.

In drawing back from the specifics of each chapter, though, there are several "big picture" issues that need to be borne in mind. Nine seem particularly pertinent.

First is the question of whether or not scale is actually a useful analytical category, either considered in terms of *size* (a horizontal measure) or *level* (a vertical one). Thus, whereas myriad geographers have long taken it to be a key geographical concept – along with others such as landscape, space, place, and location – more recently some have questioned this assumption. Indeed, Marston *et al.* (2005: 416) have even suggested "eliminat[ing] scale as a concept in human geography" and offering in its place "a different ontology, one that so flattens scale as to render the concept unnecessary." Others have maintained that scale remains a useful concept, whether it is considered in ontological or epistemological terms. Hence, Jonas (2006) has argued that it retains utility because scales are real entities with ontological presence in the landscape – economic regions, for instance, cover particular absolute spaces and end somewhere. For her part, Jones (1998) has averred that the key worth that the concept of scale provides is in epistemological (rather than ontological) terms, since it can be used by various social actors to frame particular sets of issues in certain ways – as global versus as national, for instance. Despite Marston *et al.*'s rejection, it appears that scale remains a vitally important concept through which to make sense of the unevenly developed absolute spaces of capitalism, even if we must remain vigilant to how our conception of it shapes how we engage with the material world.

Second is the issue of scale's materiality (or lack thereof). This question derives from deep philosophical differences related to matters of ontology. Specifically, whereas materialists – primarily, though not exclusively, drawing from Marx – have considered scale to have material heft, idealists – generally drawing from Kant – have considered it simply a mental contrivance. This distinction has significant implications for how landscapes are understood to be structured and the role played by scale in such structuring. In particular, a materialist view is much more likely to see spatial scale – whether in "vertical" or "horizontal" terms – as a central object of social struggle than is an idealist one. Thus, in debates over developing a Reconstructed Regional Geography in the 1980s, analysis drawing on a materialist framework generally focused upon the process of regional formation – that is to say, how the regional scale itself is made – whereas idealist approaches saw changes taking place in the economic landscape (such as industrial regions' deindustrialization) simply as the reallocation of functions between various relatively fixed spatial units: jobs simply moved from one static region to another. Indeed, idealists did not actually need to worry about having any theory of regional formation, as all that was involved for them in understanding the transformation of the geography of employment was to shift one's point of view from one preconstituted scale to another – hence, stepping back to adopt a "national" point of view would allow them to see how jobs that used to be located within one region were relocating to be within another. Put another way, materialist understandings of regional formation and transformation are deeply embedded in theorizations of the production of an unevenly developed landscape under capitalism, in which scales are themselves seen as actively reshaped by the socio-spatial processes at play. In idealist approaches, on the other hand, scales remain as discrete containers which are unimpacted by the sociospatial processes they contain, although they can be remade through some deft redrawing on the map by the regional geographer. Idealist approaches fail to see scales as social products and instead reify them. Consequently, they cannot answer the fundamental question of from whence scales such as the regional come, nor how they are transformed.

Third, emerging out of point two is the question of the relationship – if any – between different scales. Thus, although both idealist and materialist approaches might recognize that different scales appear to have been more or less important at different historical moments – the urban seemed to become more significant with the rise of industrial capitalism, for instance – approaches which view scales as materially produced entities are more likely to problematize how changes at one scale can spawn changes at others than are idealist approaches. This is because they see scales as emerging out of deep sociospatial processes which course through a capitalism that is understood to be planetary in structure and which connects the very local with the truly global. Accordingly, what goes on at one scale significantly shapes what goes on at another. By way of contrast, given that idealist approaches see scales simply as mental contrivances, there is little in this latter approach to suggest that scales are organically connected, such that what happens at one scale will necessarily have implications for what happens at another – there are, in other words, no transmission belts linking causal processes operating at one scale with what may be their material outcomes at another.

Fourth, there is the matter of the relationship between how scales are represented – as areal or as networked – and how they actually are. For many post-structuralists, at least those who follow writers such as Baudrillard (1994), this actually is not really much of an issue, as they generally argue that it is impossible to get behind the surface mani-festations of any particular representation. For them, representation _is_ reality. For materialists, however, how the world is presented to be and how it actually is are not necessarily the same thing – remembering Marx, they suggest that if there is no difference between appearance and reality then there is no need for science. Consequently, materialists insist (and I count myself amongst this group), regardless of how scales are envisioned ideationally, whether as areal or as networked, there is a material reality to them, one which must be uncovered if we are to understand the scaled nature of the world. This uncovering cannot be achieved by contemplation. Rather, it must be done by physically engaging with the world to discover its scalar structure – is it networked, is it areally scaled, or is it the case that different elements within it are scaled in different fashion?

Fifth, it is necessary to ponder how scales, as what, after Markusen (1996), we might consider somewhat "sticky" and enduring entities within the slippery fluidity of ongoing sociospatial processes, structure how such processes unfold. Thus, whereas political conflicts shape how scales are produced, how such scales are produced can also shape political conflicts. For instance, national boundaries affect the flow of capital across the economic landscape and thus how the unevenly developed geography of capitalism is produced. Likewise, struggles between unionized workers and managers over the geography of contract bargaining may result in the establishment of a uniform national wage rate in place of myriad local or regional ones, an outcome that makes it impossible for employers to play workers in different parts of a country against each other on the basis of wage rate differentials. The creation of a national scale contract, in other words, shapes the kinds of political strategies available either to management or to workers. At the same time, how scales are positioned discursively can also be shaped by, and shape, political struggles. Hence, in worker–manager conflicts over, say, wage rates or laboring conditions, firms may consciously present themselves as "global" to intimidate workers. In some cases this may be successful, and workers end up thinking it pointless to challenge management. On other occasions, though, workers may decide to build a global organization to match the apparent global scale of their employer or, perhaps ironically, they may so internalize a belief in the impossibility of challenging a firm globally that they seek to do so through interventions at the national or local level (through, say, urging national government regulation or through focusing their actions on one or two key plants within the firm's organizational structure). In turn, workers' choices will shape how scales are subsequently presented. Thus, if workers secure international union organizations' help, the firm may shift from presenting itself as a global company to a local one with important investments in the community in efforts to portray the union organizations as "outsiders" and therefore, perhaps, as less legitimate representatives of the community's interests than is the locally present firm (even though the firm itself may actually be a multi-billion dollar corporation headquartered overseas).

Sixth is the issue of how the social production of scale is linked to the broader production of the geography of capitalism (or any other social/economic system). A couple of matters arise here. First, we must recognize that scales produced under one system of social and economic life may be inherited by another but then thoroughly transformed – as with how the global was thoroughly altered in the transition from a world system based on mercantilism to one based on capitalist wage labor or how the urban was constituted differently and played different roles under European feudalism than under industrial capitalism. This means that scales have a palimpsestic quality – elements of previously produced and congealed scalar fixes may continue to shape those currently being produced even as the latter's production gradually erases them. Second, it is important to distinguish the production of space from the production of scale, for although "scale is a produced societal metric that differentiates space . . . it is not space *per se*" (Marston and Smith 2001: 615). This, in turn, raises the question of whether the same sets of processes which produce space in particular places at particular historical moments also produce the scales which give it form or whether there are different processes that do so and, if the latter, how they are connected or not with the former.

Seventh, it is important to avoid any kind of scalar fetishism wherein certain scales are associated with particular characteristics. In this regard, there has been a significant effort in recent years to destabilize a litany of dualisms that are frequently aligned with one another – global = abstract = space = powerful, whilst local = concrete = place = weak, for instance. At the same time, though, this is not to say that in specific instances certain scales do not adopt particular characteristics – it may well be that social actors organized transnationally are more powerful than are those confined to a particular city, but it may well also be the case that the opposite is true. The key, then, is that such things must be determined through investigating the world rather than through a priori assumption. Additionally, whilst some dualisms (say, weak versus strong) may serve a purpose if they are not allocated to particular scales ahead of time, others, such as the global = abstract space/local = concrete place dualism, must simply be abandoned, for if the global scale is built upon/ intimately connected to other scales (whether conceived of in network terms or in terms of how the global sits atop a hierarchy of smaller scales), how can it be any less concrete than these other, "smaller" scales? Hence, as Massey (2004: 7) has put it:

> If we sign up to the relational constitution of the world – in other words to the mutual constitution of the local and the global – then [any] counterposition between [abstract] space and [concrete] place is on shaky ground. The 'lived reality of our daily lives', invoked so often to buttress the meaningfulness of place, is in fact pretty much dispersed in its sources and its repercussions. The degree and nature of this dispersal will of course vary ... but the general proposition makes it difficult seriously to posit 'space' as the abstract outside of 'place' as lived. Where would you draw the line around 'the grounded reality of your daily life'? ... The [issue, then,] is *not* that place is not concrete, grounded, real, but rather that space – global space – *is so too.*

Eighth, scales' significances vary across time and space, for, if I might be permitted to adapt a particularly apposite turn of phrase used by Schein (1997: 662) to talk about landscapes, "[scales] are always in the process of 'becoming,' no longer reified or concretized – inert and there – but continually under scrutiny, at once manipulable and manipulated, always subject to change, and everywhere implicated in the ongoing formulation of social life." Because scales are always becoming, always subject to change, and everywhere implicated in the ongoing formulation of life, the hold that they have over particular groups of people as they live their lives and make the world around them will vary. Thus, different contemporary cultures view the body quite dissimilarly, whereas in, say, Western Europe the urban scale's importance has varied historically. This means that people may see a particular scale's integrity and potency wax and wane over the course of their lifetime living in a specific place, or they may find, as they travel from place to place, that certain scalar resolutions (such as the national) have different integrities and potencies in different parts of the world, depending upon the level of development of the social processes that produce them. There is, in other words, an unevenly developed geography to scales' production and articulation. This fact in turn can shape how social actors consider engaging in praxis to construct, either materially or ideationally, scales of social organization and what they understand themselves to be doing – in places where nation-states are strong and their national integrity unquestioned, social actors may adopt different scalar strategies from where the opposite is the case. Equally, it means that any system of scalar classification is historically and geographically situated rather than universal.

Finally, we must consider the relationship between geographical scale and the production of knowledge. There are two aspects of this consideration. First, how social organizations are scaled can dramatically shape the spatial diffusion of what is known and by whom. For example, if an organization such as a labor union has a highly vertically scaled structure, it may mean that a local branch union in one region of a country seeking information on the management practices of a multi-locational firm for whom its members work has to communicate with workers in different parts of the country via its regional and national organizations. Thus, a question is sent up the organizational hierarchy to the union's regional labor council and then onto its national headquarters before being sent to a regional labor council in a different part of the country in which the same firm has a facility for delivery to the appropriate local branch in that region. On the other hand, if the union has a flatter scalar organization, local branch unions in one region may have the authority to communicate directly with their confrères in other parts of the country. In the first instance, the boundaries of the regional divisions into which the union is partitioned serve to prevent information being transmitted across the landscape horizontally between regions – information can only flow between regions if it travels via the national headquarters. This is not the case in the second instance, however, and information may flow more readily across the landscape, avoiding the national headquarters altogether. Relatedly, as Berg (2004) has maintained, certain ideas are seen as attached to particular local or regional places whereas others are seen as either national or global or, essentially, placeless because of their universalism – pursuing "national" legislation may be seen as the purview of the union's national head-quarters whereas lobbying various state governments may be seen as the job of the appropriate regional organization.

Second, the idea that the world is scaled (or not) recursively shapes how we comprehend its nature. In other words, ideas about scale structure the knowledge we create about the scaled nature of world. This is not to argue for an uninspiring Kantianism in which any scalar order seen in the world is simply that which is imposed on it by our brain. Rather, it is to acknowledge that how we have been trained to examine the world shapes how we see it. Thus, as Sayer (1984: 51–52) has argued, although a biologist and a layperson may both look down a microscope, they are likely to see quite different things – bacteria versus tiny "blobs" in water – as a result of how each has been trained (or not) to see. Hence, what we see "is . . . not simply a function of the physical receptivity of our sense organs: it is also strongly influenced by the extent to which we take for granted and hence forget the concepts involved in perception." This recognition does not materially change what is at the end of the microscope, but it does change what is understood to be there. In the case of scale, then, believing the world to be scaled (or not) is likely to shape how we engage with it and so the kind of knowledge about its materiality that we produce. This fact suggests a need to constantly interrogate the relationship between the realm of ideas and the realm of material things, and also not to confuse the one for the other.

BIBLIOGRAPHY

Aalberts, T.E. (2002) "Multilevel governance and the future of sovereignty: A constructivist perspective," Department of Political Science Working Paper No. 04/2002, Free University, Amsterdam.

Abrahamson, M. (2004) *Global Cities*, New York: Oxford University Press.

Adelstein, R.P. (1991) "'The nation as an economic unit': Keynes, Roosevelt, and the managerial ideal," *The Journal of American History*, 78.1: 160–87.

Adey, P. (2010) *Mobility*, London: Routledge.

Agnew, J. (1999) "Regions on the mind does not equal regions of the mind," *Progress in Human Geography*, 23.1: 91–96.

—— (2005) "Sovereignty regimes: Territoriality and state authority in contemporary world politics," *Annals of the Association of American Geographers*, 95.2: 437–61.

—— (2008) "Borders on the mind: Re-framing border thinking," *Ethics and Global Politics*, 1.4: 1–17.

Agnew, J. and Corbridge, S. (1995) *Mastering Space: Hegemony, Territory and International Political Economy*, New York: Routledge.

Allen, J. and Cochrane, A. (2007) "Beyond the territorial fix: Regional assemblages, politics and power," *Regional Studies*, 41.9: 1161–75.

Althusser, L. (1965/1997) *Reading Capital*, New York: Verso.

Amariglio, J. (1988) "The body, economic discourse, and power: An economist's introduction to Foucault," *History of Political Economy*, 20.4: 583–613.

Amin, A. (1994) *Post-Fordism: A Reader*, Oxford: Blackwell.

—— (2002) "Spatialities of globalisation," *Environment and Planning A*, 34.3: 385–99.

—— (2004) "Regions unbound: Towards a new politics of place," *Geografiska Annaler B*, 86.1: 33–44.

Amin, A. and Thrift, N. (2002) *Cities: Reimagining the Urban*, Cambridge: Polity Press.

—— (2005) "Seeing the city as a site of international influence," *Harvard International Review*, 27.3 (online version).

Amin, S., Arrighi, G., Frank, A.G., and Wallerstein, I. (1982) *Dynamics of Global Crisis*, New York: Monthly Review Press.

Andersen, H.C. (1913) *Creation of a World Centre of Communication*, Paris: Philippe Renouard.

Anderson, B. (1983) *Imagined Communities: Reflections on the Origin and Spread of Nationalism*, London: Verso.

Anderson, J. (1996) "The shifting stage of politics: New medieval and postmodern territorialities?," *Environment and Planning D: Society and Space*, 14.2: 133–53.

Appadurai, A. (1996) "Sovereignty without territoriality: Notes for a postnational geography," in P. Yaeger (ed.), *The Geography of Identity*, 40–58, Ann Arbor: University of Michigan Press.

Archer, K. (1993) "Regions as social organisms: The Lamarckian characteristics of Vidal de la Blache's regional geography," *Annals of the Association of American Geographers*, 83.3: 498–514.

Arntz, K. (1999) "Landscape: A forgotten legacy," *Area*, 31.3: 297–300.

Badie, B. and Birnbaum, P. (1979) *Sociologie de l'État* [*Sociology of the State*], Paris: Grasset.

Baker, A.R.H. (2003) *Geography and History: Bridging the Divide*, Cambridge: Cambridge University Press.

Baker, O.E. (1926) "Agricultural regions of North America. Part I – The basis of classification," *Economic Geography*, 2.4: 459–93.

—— (1927a) "Agricultural regions of North America. Part II – The South," *Economic Geography*, 3.1: 50–86.

—— (1927b) "Agricultural regions of North America. Part III – The Middle Country where South and North meet," *Economic Geography*, 3.3: 309–39.

Bakunin, M.A. (1867–72/1950) *Marxism, Freedom and the State*, London: Freedom Press.

Barabási, A.-L. (2002) *Linked: The New Science of Networks*, Cambridge, MA: Perseus.

Barber, B.R. (2001) *Jihad vs. McWorld: How Globalism and Tribalism are Reshaping the World*, New York: Ballantine Books.

Barlow, M. and Slack, B. (1985) "International cities: Some geographical considerations and a case study of Montreal," *Geoforum*, 16.3: 333–45.

Bartelson, J. (2000) "Three concepts of globalization," *International Sociology*, 15.2: 180–96.

Bartky, I.R. (2000) *Selling the True Time: Nineteenth-Century Timekeeping in America*, Stanford, CA: Stanford University Press.

Bassett, T.J. (1994) "Cartography and empire building in nineteenth-century West Africa," *Geographical Review*, 84.3: 316–35.

Bassin, M. (1987) "Imperialism and the nation state in Friedrich Ratzel's political geography," *Progress in Human Geography*, 11.4: 473–95.

Batten, D.F. (1995) "Network cities: Creative urban agglomerations for the 21st century," *Urban Studies*, 32.2: 313–27.

Baudrillard, J. (1994) *Simulacra and Simulation*, Ann Arbor: University of Michigan Press.

Bauman, Z. (1998) *Globalization: The Human Consequences*, New York: Columbia University Press.

de Beauvoir, S. (1949/1953) *The Second Sex*, New York: Alfred A. Knopf.

Bell, D. and Valentine, G. (1995) *Mapping Desire: Geographies of Sexualities*, London: Routledge.

Benedict, B. (1983) *Imagined Communities: Reflections on the Origin and Spread of Nationalism*, London: Verso.

Bennett, R.J. and Wilson, A.G. (2003) "Geography applied," in R. Johnston and M. Williams (eds), *A Century of British Geography*, 463–501, Oxford: Oxford University Press.

Berdoulay, V. (1981) *La formation de l'école française de géographie (1870–1914)* [*The Formation of the French School of Geography (1870–1914)*], Paris: Bibliothèque Nationale/Comité des Travaux Historiques et Scientifiques.

Berg, L.D. (2004) "Scaling knowledge: Towards a *critical geography* of critical geographies," *Geoforum*, 35.5: 553–58.

Berry, B.J.L. (1965) "Internal structure of the city," *Law and Contemporary Problems*, 30.1: 111–19.

—— (1966) "Essays on commodity flows and the spatial structure of the Indian economy," Research Paper no. 111, Department of Geography, University of Chicago.

—— (1967) "Grouping and regionalizing: An approach to the problem using multivariate analysis," in W.L. Garrison and D.F. Marble (eds), *Quantitative Geography Part I: Economic and Cultural Topics*, 219–51, Northwestern University Studies in Geography No. 13, Evanston, IL.

Berry, B.J.L. and Garrison, W.L. (1958) "Recent developments in Central Place Theory," *Papers and Proceedings of the Regional Science Association*, 4:107–20.

Bertinetto, P.M. and Loporcaro, M. (2005) "The sound pattern of Standard Italian, as compared with the varieties spoken in Florence, Milan and Rome," *Journal of the International Phonetic Association*, 35.2: 131–51.

Bhabha, H.K. (1994) *The Location of Culture*, New York: Routledge.

Birnbaum, P. (2001) *The Idea of France*, New York: Hill and Wang.

Blaise, C. (2000) *Time Lord: Sir Sandford Fleming and the Creation of Standard Time*, New York: Pantheon Books.

Blom, A. (no date) "'La guerre fait l'Etat': Trajectoires extra-occidentales et privatisation de la violence" ["'War makes the state': Non-Western trajectories and the privatization of violence"], Working Paper No.2, Centre d'études en sciences sociales de la défense, French Ministry of Defense, Paris.

Bloustien, G. (2001) "Far from sugar and spice: Teenage girls, embodiment and representation," in B. Baron and H. Kotthoff (eds), *Gender in Interaction: Perspectives on Femininity and Masculinity in Ethnography and Discourse*, 99–136, Philadelphia, PA: John Benjamins Publishing Company.

Bluestone, B. and Harrison, B. (1982) *The Deindustrialization of America: Plant Closings, Community Abandonment, and the Dismantling of Basic Industries*, New York: Basic Books.

Blumen, O. and Kellerman, A. (1990) "Gender differences in commuting distance, residence, and employment location: Metropolitan Haifa 1972 and 1983," *Professional Geographer*, 42.1: 54–71.

Bohannan, P. and Bohannan, L. (1968) *Tiv Economy*, Evanston, IL: Northwestern University Press.

Bondi, L. and Rose, D. (2003) "Constructing gender, constructing the urban: A review of Anglo-American feminist urban geography," *Gender, Place and Culture*, 10.3: 229–45.

Bordo, S. (1987) "The Cartesian masculinization of thought," in S. Harding and J.F. O'Barr (eds), *Sex and Scientific Inquiry*, 247–64, Chicago: University of Chicago Press.

—— (1993) *Unbearable Weight: Feminism, Western Culture, and the Body*, Berkeley and Los Angeles: University of California Press.

Bourdieu, P. (1991) *Language and Symbolic Power*, Cambridge, MA: Harvard University Press.

Bowman, I. (1924) *Desert Trails of Atacama*, New York: American Geographical Society.

Braden, K.E. (1992) "Regions, Semple, and structuration," *Geographical Review*, 82.3: 237–43.

Brealey, K.G. (1995) "Mapping them 'out': Euro-Canadian cartography and the appropriation of the Nuxalk and Ts'ilhqot'in First Nations' territories, 1793–1916," *The Canadian Geographer/Le Géographe canadien*, 39.2: 140–56.

Brenner, N. (1997) "Global, fragmented, hierarchical: Henri Lefebvre's geographies of globalization," *Public Culture*, 10.1: 135–67.

—— (1998a) "Between fixity and motion: Accumulation, territorial organization and the historical geography of spatial scales," *Environment and Planning D: Society and Space*, 16.4: 459–81.

—— (1998b) "Global cities, glocal states: Global city formation and state territorial restructuring in contemporary Europe," *Review of International Political Economy*, 5.1: 1–37.

—— (2000) "The urban question as a scale question: Reflections on Henri Lefebvre, urban theory and the politics of scale," *International Journal of Urban and Regional Research*, 24.2: 361–78.

—— (2001) "The limits to scale? Methodological reflections on scalar structuration," *Progress in Human Geography*, 25.4: 591–614.

Brenner, N. and Elden, S. (2009) "Introduction. State, space, world: Lefebvre and the survival of capitalism," in N. Brenner and S. Elden (eds), *State, Space, World: Selected Essays/Henri Lefebvre*, 1–48, Minneapolis: University of Minnesota Press.

Brewer, A. (1990) *Marxist Theories of Imperialism: A Critical Survey*, 2nd edn, London: Routledge.

Browett, J. (1984) "On the necessity and inevitability of uneven spatial development under capitalism," *International Journal of Urban and Regional Research*, 8.2: 155–76.

Brown, J.C. and Purcell, M. (2005) "There's nothing inherent about scale: Political ecology, the local trap, and the politics of development in the Brazilian Amazon," *Geoforum*, 36.5: 607–24.

Brown, M. (1995) "Sex, scale and the 'new urban politics': HIV-prevention strategies from Yaletown, Vancouver," in D. Bell and G. Valentine (eds), *Mapping Desire: Geographies of Sexualities*, 245–63, New York: Routledge.

—— (2000) *Closet Space: Geographies of Metaphor from the Body to the Globe*, New York: Routledge.

Brown, P. (1998) "Biology and the social construction of the 'race' concept," in J. Ferrante and P. Brown (eds), *The Social Construction of Race and Ethnicity in the United States*, 131–38, New York: Longman.

Bryan, L. and Farrell, D. (1996) *Market Unbound: Unleashing Global Capitalism*, New York: Wiley.

Buisseret, D. (ed.) (1992) *Monarchs, Ministers, and Maps: The Emergence of Cartography as a Tool of Government in Early Modern Europe*, Chicago: University of Chicago Press.

Bunge, W. (1962) *Theoretical Geography*, Lund: Lund Studies in Geography, Department of Geography, University of Lund, Sweden.

Burgel, G., Burgel, G. and Dezes, M.G. (1987) "An interview with Henri Lefebvre," *Environment and Planning D: Society and Space*, 5.1: 27–38.

Burkholder, M.A. and Johnson, L.L. (2003) *Colonial Latin America*, 5th edn, Oxford: Oxford University Press.

Burton, I. (1963) "The quantitative revolution and theoretical geography," *Canadian Geographer*, 7.4: 151–62.

Butler, J. (1990) *Gender Trouble: Feminism and the Subversion of Identity*, New York: Routledge.

Buttimer, A. (1971) *Society and Milieu in the French Geographic Tradition*, Chicago: A.A.G. Monograph No. 6, Rand McNally.

Butzer, K.W. (1989) "Hartshorne, Hettner, and *The Nature of Geography*," in J.N. Entrikin and S.D. Brunn (eds), *Reflections of Richard Hartshorne's The Nature of Geography*, 35–52, Washington, DC: Occasional Publications of the Association of American Geographers.

Camagni, R.P. (1993) "From city hierarchy to city network: Reflections about an emerging paradigm," in T.R. Lakshmanan and P. Nijkamp (eds), *Structure and Change in the Space Economy: Festschrift in Honor of Martin J. Beckmann*, 66–87, Berlin: Springer-Verlag.

Camagni, R.P. and Salone, C. (1993) "Network urban structures in Northern Italy: Elements for a theoretical framework," *Urban Studies*, 30.6: 1053–64.

Campbell, C.S. (1994) "The second Nature of Geography: Hartshorne as humanist," *Professional Geographer*, 46.4: 411–17.

Cañeque, A. (2004) *The King's Living Image: The Culture and Politics of Viceregal Power in Colonial Mexico*, New York: Routledge.

Carpenter, E. and McLuhan, M. (eds) (1960) *Explorations in Communication: An Anthology*, Boston: Beacon Press.

Carr, M. (1997) *New Patterns: Process and Change in Human Geography*, Cheltenham, UK: Nelson Thornes.

Carson, C.S. (1975) "The history of the United States national income and product accounts: The development of an analytical tool," *Review of Income and Wealth*, 21.2: 153–81.

Castells, M. (1977) *The Urban Question: A Marxist Approach*, Cambridge, MA: MIT Press.

—— (1978) *City, Class and Power*, Basingstoke, UK: Macmillan.

—— (1983) *The City and the Grassroots: A Cross-Cultural Theory of Urban Social Movements*, Berkeley and Los Angeles: University of California Press.

—— (1996) *The Rise of the Network Society (Volume 1 – The Information Age: Economy, Society and Culture)*, Oxford: Blackwell.

—— (1999/2002) "The culture of cities in the information age," in I. Susser (ed.) *The Castells Reader on Cities and Social Theory*, 367–89, Oxford: Blackwell.

Castells, M. (ed.) (2004) *The Network Society: A Cross-Cultural Perspective*, Cheltenham, UK: Edward Elgar.

Castells, M. and Cardoso, G. (eds) (2006) *The Network Society: From Knowledge to Policy*, Washington, DC: Johns Hopkins Center for Transatlantic Relations.

Castree, N., Featherstone, D. and Herod, A. (2007) "Contrapuntal geographies: The politics of organising across socio-spatial difference," in K. Cox, M. Low and J. Robinson (eds) *The SAGE Handbook of Political Geography*, 305–21, London: Sage.

Catchpowle, L., Cooper, C. and Wright, A. (2004) "Capitalism, states and accounting," *Critical Perspectives on Accounting*, 15.8: 1037–58.

Chang, K.A. and Ling, L.H.M. (2000) "Globalization and its intimate other: Filipina domestic workers in Hong Kong," in M.H. Marchand and A.S. Runyan (eds), *Gender and Global Restructuring: Sightings, Sites and Resistances*, 27–43, London: Routledge.

Chapura, M. (2009) "Scale, causality, complexity and emergence: Rethinking scale's ontological significance," *Transactions of the Institute of British Geographers*, New Series, 34.4: 462–74.

Christaller, W. (1933/1966) *Central Places in Southern Germany* [Originally published as *Die zentralen Orte in Süddeutschland*], Englewood Cliffs, NJ: Prentice-Hall.

Christiansen, T. (1996) "Reconstructing European space: From territorial politics to multi-level governance," European University Institute Working Paper RSC No. 53/96, Florence: European University Institute.

Claval, P. (1984) "France," in R.J. Johnston and P. Claval (eds), *Geography Since the Second World War: An International Survey*, 15–41, London: Croom Helm.

—— (2008) "Les espaces de l'économie" ("Spaces of the economy"), *Annales de géographie*, 664: 3–22.

Claval, P. and Thompson, I. (1998) *An Introduction to Regional Geography*, Oxford: Blackwell.

Clements, F.E. (1916) *Plant Succession: An Analysis of the Development of Vegetation*, Washington, DC: Carnegie Institution of Washington.

Clout, H. (2003a) "In the shadow of Vidal de la Blache: Letters to Albert Demangeon and the social dynamics of French geography in the early twentieth century," *Journal of Historical Geography*, 29.3: 336–55.

—— (2003b) "Place description, regional geography and area studies: The chorographic inheritance," in R.J. Johnston and M. Williams (eds), *A Century of British Geography*, 247–74, Oxford: Oxford University Press.

Coker, F.W. (1910) "Organismic theories of the state," Ph.D. thesis, Faculty of Political Science, Columbia University, New York City.

Collinge, C. (2005) "The *différance* between society and space: Nested scales and the returns of spatial fetishism," *Environment and Planning D: Society and Space*, 23.2: 189–206.

—— (2006) "Flat ontology and the deconstruction of scale: A response to Marston, Jones and Woodward," *Transactions of the Institute of British Geographers*, New Series, 31.2: 244–51.

Conway, J. (2008) "Geographies of transnational feminisms: The politics of place and scale in the World March of Women," *Social Politics: International Studies in Gender, State and Society*, 15.2: 207–31.

Conzen, M.P. (1975) "A transport interpretation of the growth of urban regions: An American example," *Journal of Historical Geography*, 1.4: 361–82.

Cooke, P. (1985) "Class practices as regional markers: A contribution to labour geography," in D. Gregory and J. Urry (eds), *Social Relations and Spatial Structures*, 213–41, New York: St. Martin's Press.

—— (1987) "Clinical inference and geographic theory," *Antipode*, 19.1: 69–78.

Cooke, P. (ed.) (1989) *Localities: The Changing Face of Urban Britain*, London: Unwin Hyman.

Cooper, R. (1984) "A note on the biologic concept of race and its application in epidemiological research," *American Heart Journal*, 108.3: 715–23.

Cosgrove, D. (1994) "Contested global visions: *One-World*, *Whole-Earth*, and the Apollo space photographs," *Annals of the Association of American Geographers*, 84.2: 270–94.

—— (2001) *Apollo's Eye: A Cartographic Genealogy of the Earth in the Western Imagination*, Baltimore: The Johns Hopkins University Press.

—— (2003) "Globalism and tolerance in early modern geography," *Annals of the Association of American Geographers*, 93.4: 852–70.

Cox, K.R. (1998) "Spaces of dependence, spaces of engagement and the politics of scale, or: Looking for local politics," *Political Geography*, 17.1: 1–23.

Cox, K.R. and Mair, A. (1988) "Locality and community in the politics of local economic development," *Annals of the Association of American Geographers*, 78.2: 307–25.

—— (1991) "From localised social structures to localities as agents," *Environment and Planning A*, 23.2: 197–213.

Crone, G.R. (1951) *Modern Geographers: An Outline of Progress in Geography Since 1800 A.D.*, London: Royal Geographical Society.

Crump, J. and Merrett, C. (1998) "Scales of struggle: Economic restructuring in the US Midwest," *Annals of the Association of American Geographers*, 88.3: 496–515.

Dale, E.M. (2006) "Hegel, evil, and the end of history," in A. MacLachlan and I. Torsen (eds), *History and Judgement*, Vienna: Institut für die

Wissenschaften vom Menschen (IWM) Junior Visiting Fellows' Conferences, Vol. 21.

Darby, H.C. (1940a) *The Draining of the Fens*, Cambridge: Cambridge University Press.

—— (1940b) *The Medieval Fenland*, Cambridge: Cambridge University Press.

Davis, K. (2007) *The Making of Our Bodies, Ourselves: How Feminism Travels Across Borders*, Durham, NC: Duke University Press.

Davis, L.J. (1995) *Enforcing Normalcy: Disability, Deafness, and the Body*, New York: Verso.

Delaney, D. and Leitner, H. (1997) "The political construction of scale," *Political Geography*, 16.2: 93–97.

Delgado, R. and Stefancic, J. (2001) *Critical Race Theory: An Introduction*, New York: New York University Press.

Derrida, J. (1981) *Dissemination*, London: Athlone Press.

Descartes, R. (1641/1996) *Discourse on the Method; and, Meditations on First Philosophy* (edited by D. Weissman), New Haven, CT: Yale University Press.

Dickinson, R.E. (1939) "Landscape and society," *Scottish Geographical Magazine*, 55.1: 1–14.

—— (1976) *Regional Concept: The Anglo-American Leaders*, London: Routledge and Kegan Paul.

Dikshit, R.D. (1999) *Political Geography*, 3rd edn, New Delhi: Tata McGraw-Hill.

Dirlik, A. (1999) "Place-based imagination: Globalism and the politics of place," unpublished manuscript, Department of History, Duke University, Durham, NC.

Dodge, R.E. (1935) "Response to Platt, R.S. (1935) Field approach to regions," *Annals of the Association of American Geographers*, 25.3: 172–74.

Doel, M. (1999) *Poststructuralist Geographies: The Diabolical Art of Spatial Science*, Edinburgh: Edinburgh University Press.

Dryer, C.R. (1912) "Regional geography," *The Journal of Geography*, 11.3: 73–75.

—— (1915) "Natural economic regions," *Annals of the Association of American Geographers*, 5: 121–25.

Duncan, J. (1978) "Men without property: The tramp's classification and use of urban space," *Antipode* 10.1: 24–34.

—— (1980) "The superorganic in American cultural geography," *Annals of the Association of American Geographers*, 70.2: 181–98.

Durkheim, E. (1893/1947) *The Division of Labor in Society*, Glencoe, IL: Free Press.

Eisler, R. (2000) *Tomorrow's Children: A Blueprint for Partnership Education in the 21st Century*, Boulder, CO: Westview Press.

Elazar, D.J. (2004) "Jerusalem: The ideal city of the Bible," *Jewish Political Studies Review*, 16.1–2 (online publication; no page numbers).

Elden, S. (2005) "Missing the point: Globalization, deterritorialization and the space of the world," *Transactions of the Institute of British Geographers*, New Series, 30.1: 8–19.

Elkins, T.H. (1989) "Human and regional geography in the German-speaking lands in the first forty years of the twentieth century," in J.N. Entrikin and S.D. Brunn (eds), *Reflections of Richard Hartshorne's The Nature of Geography*, 17–34, Washington, DC: Occasional Publications of the Association of American Geographers.

Engels, F. (1844/1993) *The Condition of the Working Class in England*, Oxford: Oxford University Press.

Entriken, J.N. (1996) "Place and region 2," *Progress in Human Geography*, 20.2: 215–21.

Eschle, C. (2004) "Feminist studies of globalisation: Beyond gender, beyond economism?," *Global Society*, 18.2: 97–125.

Falk, P. (1994) *The Consuming Body*, London: Sage.

Fenneman, N.M. (1914) "Physiographical boundaries within the United States," *Annals of the Association of American Geographers*, 4: 84–134.

Fèvre, J. and Hauser, H. (1909) *Régions et Pays de France* [*Regions and Lands of France*], Paris: Félix Alcan.

Fincher, R. (1990) "Women in the city: Feminist analyses of urban geography," *Australian Geographical Studies*, 28:1, 29–37.

Finkelstein, A. (2000) *Harmony and the Balance: An Intellectual History of Seventeenth-Century English Economic Thought*, Ann Arbor: University of Michigan Press.

Firestone, S. (1970) *The Dialectic of Sex: The Case for Feminist Revolution*, New York: William Morrow.

Fleure, H.J. (1919) "Human regions," *Scottish Geographical Magazine*, 35.3: 94–105.

Foord, J. and Gregson, N. (1986) "Patriarchy: Towards a reconceptualization," *Antipode*, 18.2: 186–211.

Forest, B. (1995) "West Hollywood as symbol: The significance of place in the construction of a gay identity," *Environment and Planning D: Society and Space*, 13.2: 133–57.

Foucault, M. (1975/1977) *Discipline and Punish: The Birth of the Prison*, New York: Pantheon Books.

—— (1978/1997) "Security, territory, and population," in P. Rabinow (ed.), *Michel Foucault – Ethics: Subjectivity and Truth*, 67–71, New York: The New Press.

—— (1978/2007) "Spaces of security: The example of the town. Lecture of 11th January 1978," *Political Geography*, 26.1, 48–56.

—— (1986) "Of other spaces," *Diacritics*, 16.1, 22–27.

Freeman, C. (2001) "Is local: global as feminine: masculine? Rethinking the gender of globalization," *Signs: Journal of Women in Culture and Society*, 26.4, 1007–37.

Freud, S. (1927) *The Ego and the Id*, London: Hogarth Press.

Friedmann, J. (1986) "The World City hypothesis," *Development and Change*, 17.1: 69–83.

—— (1995) "Where we stand: A decade of world city research," in P.L. Knox and P.J. Taylor (eds), *World Cities in a World-System*, 21–47, Cambridge: Cambridge University Press.

Friedmann, J. and Wolff, G. (1982) "World city formation: An agenda for research and action," *International Journal of Urban and Regional Research*, 6.3: 309–44.

Fukuyama, F. (1992) *The End of History and the Last Man*, New York: Free Press.

Gabardi, W. (2000) *Negotiating Postmodernism*, Minneapolis: University of Minnesota Press.

Galison, P. (2003) *Einstein's Clocks, Poincaré's Maps: Empires of Time*, New York: W.W. Norton.

Geddes, P. (1915/1950) *Cities in Evolution*, New York: Oxford University Press.

Gerschenkron, A. (1974) "Figures of speech in social sciences," *Proceedings, American Philosophical Society*, 118.5: 431–48.

Gibson, K. (2001) "Regional subjection and becoming," *Environment and Planning D*, 19.6: 639–67.

Gibson-Graham, J.K. (1996) *The End of Capitalism (As We Knew it): A Feminist Critique of Political Economy*, Oxford: Blackwell.

—— (2002) "Beyond global vs. local: Economic politics outside the binary frame," in A. Herod and M.W. Wright (eds), *Geographies of Power: Placing Scale*, 25–60, Oxford: Blackwell.

Giddens, A. (1981) *A Contemporary Critique of Historical Materialism*, London: MacMillan.

—— (1984) *The Constitution of Society: Outline of the Theory of Structuration*, Berkeley and Los Angeles: University of California Press.

—— (1987) *The Nation-State and Violence: Volume Two of a Contemporary Critique of Historical Materialism*, Berkeley and Los Angeles: University of California Press.

Gilbert, A. (1988) "The new regional geography in English and French-speaking countries," *Progress in Human Geography*, 12.2: 208–28.

Godlewska, A. (1995) "Map, text and image – The mentality of enlightened conquerors: A new look at the *Description de l'Egypte*," *Transactions of the Institute of British Geographers*, New Series, 20.1: 5–28.

Goldstone, J.A. (1991) "States making wars making states making wars . . .," *Contemporary Sociology*, 20.2: 176–78.

Goswami, M. (2004) *Producing India: From Colonial Economy to National Space*. Chicago: University of Chicago Press.

Gould, S.J. (1987) *Time's Arrow, Time's Cycle: Myth and Metaphor in the Discovery of Geological Time*, Cambridge, MA: Harvard University Press.

Goudzwaard, B. (2001) *Globalization and the Kingdom of God*, Grand Rapids, MI: Baker Books.

Gray, J. (1998) *False Dawn: The Delusions of Global Capitalism*, New York: New Press.

Green, A.E. (1988) "The North-South divide in Great Britain: An examination of the evidence," *Transactions of the Institute of British Geographers*, New Series, 13.2: 179–98.

Griffin, S. (1978) *Woman and Nature: The Roaring Inside Her*, New York: Harper and Row.

Grigg, D. (1965) "The logic of regional systems," *Annals of the Association of American Geographers*, 55.3: 465–91.

—— (1967) "Regions, models and classes," in R.J. Chorley and P. Haggett (eds), *Models in Geography*, 461–509, London: Methuen.

Grosz, E.A. (1994) *Volatile Bodies: Toward a Corporeal Feminism*, Bloomington: Indiana University Press.

Guenther, K. (2006) "'A bastion of sanity in a crazy world': A local feminist movement and the reconstitution of scale, space, and place in an Eastern German city," *Social Politics: International Studies in Gender, State and Society*, 13.4: 551–75.

Haggett, P. (1966) *Locational Analysis in Human Geography*, New York: St. Martin's Press.

—— (1972/1975) *Geography: A Modern Synthesis*, 2nd edn, New York: Harper and Row.

—— (1990) *The Geographer's Art*, Oxford: Blackwell.

Haggett, P. and Chorley, R. (1969) *Network Analysis in Geography*, London: Edward Arnold.

Hall, P.G. (1966) *The World Cities*, London: Weidenfeld and Nicolson.

Hall, W.W. Jr. and Hite, J.C. (1970) "The use of central place theory and gravity-flow analysis to delineate economic areas," *Southern Journal of Agricultural Economics*, 2.1: 147–53.

Hansen, M.H. (1995) "Kome: A study in how the Greeks designated and classified settlements which were not *Poleis*," in M.H. Hansen and K. Raaflaub (eds), *Studies in the Ancient Greek Polis*, 45–81, Stuttgart: Franz Steiner Verlag.

Hanson, S. and Pratt, G. (1995) *Gender, Work, and Space*, New York: Routledge.

Haraway, D. (1989) *Primate Visions: Gender, Race, and Nature in the World of Modern Science*, London: Routledge.

—— (1991) *Simians, Cyborgs and Women: The Reinvention of Nature*, London: Free Association Books.

Harloe, M., Pickvance, C.G. and Urry, J. (eds) (1990) *Place, Policy and Politics: Do Localities Matter?*, London: Unwin Hyman.

Harris, C.D. and Ullman, E.L. (1945) "The nature of cities," *Annals of the American Academy of Political and Social Science*, 242 (Nov): 7–17.

Harris, J.G. (2003) "Stephen Greenblatt's 'X'-Files: The rhetoric of containment and invasive disease in 'Invisible Bullets' and 'The Sources of Soviet Conduct'," in P.C. Herman (ed.), *Historicizing Theory*, 137–57, Albany, NY: SUNY Press.

Hart, J.F. (1982) "The highest form of the geographer's art," *Annals of the Association of American Geographers*, 72.1: 1–29.

Hartshorne, R. (1935a) "Recent developments in political geography, I," *American Political Science Review*, 29.5: 785–804.

—— (1935b) "Recent developments in political geography, II," *American Political Science Review*, 29.6: 943–66.

—— (1939) *The Nature of Geography: A Critical Survey of Current Thought in the Light of the Past*, Lancaster, PA: Association of American Geographers (Fourth Printing, 1951).

—— (1959) *Perspective on The Nature of Geography*, Chicago: Rand McNally.
Hartsock, N. (1983) "The Feminist Standpoint," in S. Harding and M.B. Hintikka (eds), *Discovering Reality: Feminist Perspectives on Epistemology, Metaphysics, Methodology and Philosophy of Science*, 283–310, Boston: Reidel Publishing.
Harvey, D. (1969) *Explanation in Geography*, New York: St Martin's Press.
—— (1973) *Social Justice and the City*, Baltimore: Johns Hopkins University Press.
—— (1976) "Labor, capital, and class struggle around the built environment in advanced capitalist societies," *Politics and Society*, 6.3: 265–95.
—— (1982) *The Limits to Capital*, Oxford: Basil Blackwell.
—— (1983) "The urban process under capitalism: A framework for analysis," in R. Lake (ed.), *Readings in Urban Analysis*, 197–227, New Brunswick, NJ: Center for Urban Policy Research, Rutgers University.
—— (1985) "The geopolitics of capitalism," in D. Gregory and J. Urry (eds), *Social Relations and Spatial Structures*, 128–63, New York: St. Martin's Press.
—— (1989a) *The Urban Experience*, Baltimore: Johns Hopkins University Press.
—— (1989b) *The Condition of Postmodernity: An Enquiry into the Origins of Cultural Change*, Oxford: Blackwell.
—— (2000) *Spaces of Hope*, Berkeley and Los Angeles: University of California Press.
Harvey, F., and Wardenga, U. (1998) "The Hettner-Hartshorne connection: Reconsidering the process of reception and transformation of a geographic concept," *Finisterra*, 33.65: 131–40.
—— (2006) "Richard Hartshorne's adaptation of Alfred Hettner's system of geography," *Journal of Historical Geography*, 32.2: 422–40.
Hayden, D. (1980) "What would a non-sexist city be like? Speculations on housing, urban design, and human work," *Signs: Journal of Women in Culture and Society*, 5.S3: S170-S187.
Hayduk, R. (2006) *Democracy for All: Restoring Immigrant Voting Rights in the United States*, New York: Routledge.
Hayes, A.W. (2001) *Principles and Methods of Toxicology*, 4th edn, New York: CRC Press.
Heffernan, M. (2000) *"Fin de siècle, fin du monde?*: On the origins of European geopolitics, 1890–1920," in K. Dodds and D. Atkinson (eds), *Geopolitical Traditions: A Century of Geopolitical Thought*, 27–51, New York: Routledge.
Hegel, G.W.F. (1837/2004) *The Philosophy of History*, New York: Dover.
Hepple, L.W. (1992) "Metaphor, geopolitical discourse and the military in South America," in T. Barnes and J. Duncan (eds), *Writing Worlds: Discourse, Text and Metaphor in the Representation of Landscapes*, 136–54, London: Routledge.
—— (2004) "South American heartland: The Charcas, Latin American geopolitics and global strategies," *The Geographical Journal*, 170.4: 359–67.
Herb, G.H. (1997) *Under the Map of Germany: Nationalism and Propaganda 1918–1945*, New York: Routledge.
Herbertson, A.J. (1905) "The major natural regions: An essay in systematic geography," *The Geographical Journal*, 25.3: 300–310.
Herod, A. (1991) "The production of scale in United States labour relations," *Area*, 23.1, 82–88.

—— (1994) "On workers' theoretical (in)visibility in the writing of critical urban geography: A comradely critique," *Urban Geography*, 15.7, 681–93.

—— (1997a) "Notes on a spatialized labour politics: Scale and the political geography of dual unionism in the US longshore industry," in R. Lee and J. Wills (eds), *Geographies of Economies*, 186–96, London: Edward Arnold.

—— (1997b) "Labor's spatial praxis and the geography of contract bargaining in the US east coast longshore industry, 1953–89," *Political Geography*, 16.2, 145–69.

—— (1997c) "Labor as an agent of globalization and as a global agent," in K. Cox (ed.), *Spaces of Globalization: Reasserting the Power of the Local*, 167–200, New York: Guilford.

—— (2000) "Implications of Just-in-Time production for union strategy: Lessons from the 1998 General Motors-United Auto Workers dispute," *Annals of the Association of American Geographers*, 90: 521–47.

—— (2001) *Labor Geographies: Workers and the Landscapes of Capitalism*, New York: Guilford.

—— (2008) "Scale: The local and the global," in S. Holloway, S. Rice, G. Valentine and N. Clifford (eds), *Key Concepts in Geography*, 2nd edn, 217–35, London: Sage.

—— (2009) *Geographies of Globalization: A Critical Introduction*, Oxford: Blackwell.

Herod, A. and Wright, M.W. (2002a) "Placing scale: An introduction," in A. Herod and M.W. Wright (eds), *Geographies of Power: Placing Scale*, 1–14, Oxford: Blackwell.

—— (eds) (2002b) *Geographies of Power: Placing Scale*, Oxford: Blackwell.

Hertwig, O. (1922) *Der Staat als Organismus: Gedanken zur Entwicklung der Menschheit* [*The State as Organism: Thoughts on the Development of Mankind*], Jena: G. Fischer.

Hintze, O. (1906/1975) "Military organization and the organization of the state," in F. Gilbert (ed.), *The Historical Essays of Otto Hintze*, 178–215, New York: Oxford University Press.

Hobbes, T. (1651/2002) *Leviathan*, Toronto: Broadview Press.

Hobden, S. (2001) "You can choose your sociology but you can't choose your relations: Tilly, Mann and relational sociology," *Review of International Studies*, 27.2: 281–86.

Hobsbawm, E.J. (1990) *Nations and Nationalism Since 1780: Programme, Myth, Reality*, Cambridge: Cambridge University Press.

Hoefle, S.W. (2006) "Eliminating scale and killing the goose that laid the golden egg?," *Transactions of the Institute of British Geographers*, New Series, 31.2: 238–43.

Hoekveld, G.A. (1990) "Regional geography must adapt to new realities," in R.J. Johnston, J. Hauer and G.A. Hoekveld (eds), *Regional Geography: Current Developments and Future Prospects*, 11–31, London: Routledge.

Hohenberg, P.M. and Lees, L.H. (1985) *The Making of Urban Europe, 1000–1950*, Cambridge, MA: Harvard University Press.

Hollingsworth, R.J. (1998) "Territoriality in modern societies: The spatial and institutional nestedness of national economies," in S. Immerfall (ed.), *Territoriality in the Globalizing Society: One Place or None?*, 17–37, Berlin: Springer-Verlag.

hooks, bell (1990) *Yearning: Race, Gender, and Cultural Politics*, Boston: South End Press.

Hottes, R. (1983) "Walter Christaller," *Annals of the Association of American Geographers*, 73.1: 51–54.

Howitt, R. (1998) "Scale as relation: Musical metaphors of geographical scale," *Area* 30.1: 49–58.

Hoyt, H. (1939) *The Structure and Growth of Residential Neighborhoods in American Cities*, Washington, DC: Federal Housing Administration.

Hudson, R. (2000) "One Europe or many? Reflections on becoming European," *Transactions of the Institute of British Geographers*, New Series, 25.4: 409–26.

Hughes, B. (2004) "Disability and the body," in J. Swain, C. Barnes, S. French and C. Thomas (eds), *Disabling Barriers: Enabling Environments*, 2nd edn, 63–68, Thousand Oaks, CA: Sage.

Hughes, J. (2004) *Citizen Cyborg: Why Democratic Societies Must Respond to the Redesigned Human of the Future*, Boulder, CO: Westview Press.

Ignatiev, N. (1995) *How the Irish Became White*, New York: Routledge.

Isard, W. (1956) *Location and Space-Economy*, Cambridge, MA: MIT Press.

Isin, E.F. (2001) "Istanbul's conflicting paths to citizenship: Islamization and globalization," in A.J. Scott (ed.), *Global City-Regions: Trends, Theory, Policy*, 349–68, Oxford: Oxford University Press.

Jack, S. (1988) "Cultural transmission: Science and society to 1850," in R.W. Home (ed.), *Australian Science in the Making*, 45–66, Melbourne: Cambridge University Press.

James, P.E. (1934) "The terminology of regional description," *Annals of the Association of American Geographers*, 24: 78–92.

—— (1952) "Toward a further understanding of the regional concept," *Annals of the Association of American Geographers*, 42.3: 195–222.

—— (1954) "Introduction: The field of geography," in P.E. James and C.F. Jones (eds) *American Geography: Inventory and Prospect*, 3–18, Syracuse, NY: Syracuse University Press

Jessop, B. (1982) *The Capitalist State: Marxist Theories and Methods*, Oxford: Blackwell.

Jessop, B. (2008) *State Power: A Strategic-Relational Approach*, Cambridge: Polity.

—— (2009) "Avoiding traps, rescaling states, governing Europe," in R. Keil and R. Mahon (eds), *Leviathan Undone? Towards a Political Economy of Scale*, 87–104, Vancouver: University of British Columbia Press.

Jessop, B., Brenner, N. and Jones, M. (2008) "Theorizing sociospatial relations," *Environment and Planning D: Society and Space*, 26.3: 389–401.

Joerg, W.L.G. (1914) "The subdivision of North America into natural regions: A preliminary inquiry," *Annals of the Association of American Geographers*, 4: 55–83.

John of Salisbury (1159/1990) *Policraticus: Of the Frivolities of Courtiers and the Footprints of Philosophers* (translated by Cary Nederman), Cambridge: Cambridge University Press.

Johnston, R.J. (1973) *Spatial Structures: Introducing the Study of Spatial Systems in Human Geography*, London: Methuen.

—— (2003) "Order in space: Geography as a discipline in distance," in R.J. Johnston and M. Williams (eds), *A Century of British Geography*, 303–45, Oxford: Oxford University Press.

Johnston, R.J., Pattie, C.J. and Allsopp, J.G. (1988) *A Nation Dividing?: The Electoral Map of Great Britain, 1979–1987*, London: Longman.

Johnston-Anumonwo, I., and Sultana, S. (2006) "Race, location and access to employment in Buffalo, N.Y.," in J.W. Frazier and E. Tettey-Fio (eds), *Race, Ethnicity, and Place in a Changing America*, 119–30, Binghamton, NY: Global Academic Publishing.

Jonas, A.E.G. (1994) "The scale politics of spatiality," *Environment and Planning D: Society and Space*, 12: 257–64.

—— (2006) "Pro scale: Further reflections on the 'scale debate' in human geography," *Transactions of the Institute of British Geographers*, New Series, 31.3: 399–406.

Jones, J.P. III, Woodward, K. and Marston, S.A. (2007) "Situating flatness," *Transactions of the Institute of British Geographers*, New Series, 32.2: 264–76.

Jones, K.T. (1998) "Scale as epistemology," *Political Geography*, 17.1: 25–28.

Judd, D.R. (1998) "The case of the missing scales: A commentary on Cox," *Political Geography*, 17: 29–34.

Juillard, É. (1962) "La région: Essai de définition" ["The region: A definitional essay"], *Annales de géographie*, 387: 483–99.

Kaiser, R. and Nikiforova, E. (2008) "The performativity of scale: The social construction of scale effects in Narva, Estonia," *Environment and Planning D: Society and Space*, 26.3: 537–62.

Kant, I. (1781/2007) *Critique of Pure Reason*, London: Penguin.

Kantola, J. (2006) *Feminists Theorize the State*, New York: Palgrave Macmillan.

Käsler, D. (1988) *Max Weber: An Introduction to His Life and Work*, Chicago: University of Chicago Press.

Katz, C. (2004) *Growing Up Global: Economic Restructuring and Children's Everyday Lives*, Minneapolis: University of Minnesota Press.

Kimble, G. (1951) "The inadequacy of the regional concept," in L. Dudley Stamp and S.W. Wooldridge (eds), *London Essays in Geography*, 151–74, Cambridge, MA: Harvard University Press.

Kingsbury, P. (2004) "Psychoanalytical approaches," in J.S. Duncan, N.C. Johnson and R.H. Schein (eds), *A Companion to Cultural Geography*, 108–20, Malden, MA: Blackwell.

Kirby, A. (2002) "Popular culture, academic discourse, and the incongruities of scale," in A. Herod and M.W. Wright (eds), *Geographies of Power: Placing Scale*, 171–91, Oxford: Blackwell.

Kitchin, R. and Wilton, R. (2003) "Disability activism and the politics of scale," *The Canadian Geographer/Le Géographe canadien*, 47.2: 97–115.

Knopp, L. (1992) "Sexuality and the spatial dynamics of capitalism," *Environment and Planning D: Society and Space*, 10.6: 651–69.

Kojève, A. (1947) *Introduction à la Lecture de Hegel [Introduction to the Reading of Hegel]*, Paris: Gallimard.

Komlos, J. (1990) "Height and social status in eighteenth-century Germany," *Journal of Interdisciplinary History*, 20.4: 607–21.

Kraidy, M.M. (2005) *Hybridity, or the Cultural Logic of Globalization*, Philadelphia, PA: Temple University Press.

Krasner, S.D. (1999) *Sovereignty: Organized Hypocrisy*, Princeton, NJ: Princeton University Press.

Kuhn, T.S. (1962) *The Structure of Scientific Revolutions*, Chicago: University of Chicago Press.

Lacan, J. (1977) *Écrits: A Selection*, New York: Norton.

Lagendijk, A. (2007) "The accident of the region: A strategic relational perspective on the construction of the region's significance," *Regional Studies*, 41.9: 1193–1208.

Laqueur, T.W. (1990) *Making Sex: Body and Gender from the Greeks to Freud*, Cambridge, MA: Harvard University Press.

Latham, A. (2002) "Retheorizing the scale of globalization: Topologies, actor-networks, and cosmopolitanism," in A. Herod and M.W. Wright (eds), *Geographies of Power: Placing Scale*, 115–44, Oxford: Blackwell.

Latour, B. (1993) *We Have Never Been Modern* (translated by C. Porter), Cambridge, MA: Harvard University Press.

—— (1996) "On actor-network theory: A few clarifications," *Soziale Welt*, 47: 369–81.

Lauria, M. and Knopp, L. (1985) "Toward an analysis of the role of gay communities in the urban renaissance," *Urban Geography*, 6.2: 152–69.

Lefebvre, H. (1970/2003) *The Urban Revolution*, Minneapolis: University of Minnesota Press.

—— (1973/1976) *The Survival of Capitalism: Reproduction of the Relations of Production*, London: St. Martin's Press.

—— (1974/1991) *The Production of Space*, Oxford: Basil Blackwell.

—— (1976) *De l'Etat (Tome 2): Théorie Marxiste de l'Etat de Hegel à Mao [On the State (Volume 2): Marxist Theory of the State from Hegel to Mao]*, Paris: Union Générale d'Editions "10/18" Collection.

—— (1978/2009) "Space and the state," in N. Brenner and S. Elden (eds), *State, Space, World: Selected Essays/Henri Lefebvre*, 223–53, Minneapolis: University of Minnesota Press.

LeGates, R.T. (2003) "How to study cities," in R.T. LeGates and F. Stout (eds), *The City Reader*, 9–18, New York: Routledge.

Leitner, H. and Sheppard, E. (2002) "'The city is dead, long live the net': Harnessing European interurban networks for a neoliberal agenda," *Antipode* 34.3: 495–518.

Leitner, H., Sheppard, E. and Sziarto. K.M. (2008) "The spatialities of contentious politics," *Transactions of the Institute of British Geographers*, New Series, 33.2: 157–72.

Leitner, H., Pavlik, C. and Sheppard, E. (2002) "Networks, governance, and the politics of scale: Inter-urban networks and the European Union," in A. Herod and M.W. Wright (eds), *Geographies of Power: Placing Scale*, 274–303, Oxford: Blackwell.

Lenin, V.I. (1902/1988) *What is to be Done?*, New York: Penguin Books.

—— (1917/1932) *State and Revolution*, New York: International Publishers.

—— (1926/1939) *Imperialism: The Highest Stage of Capitalism*, New York: International Publishers.

Lewis, P.F. (1985) "Beyond description," *Annals of the Association of American Geographers*, 75.4: 465–78.

Ley, D. and Samuels, M.S. (eds) (1978) *Humanistic Geography: Prospects and Problems*, London: Croom Helm.

Lichtenberger, E. (1979) "The impact of political systems upon geography: The case of the Federal Republic of Germany and the German Democratic Republic," *Professional Geographer*, 31.2: 201–11.

Linehan, D. (2003) "Regional survey and the economic geographies of Britain 1930–39," *Transactions of the Institute of British Geographers*, New Series, 28.1: 96–122.

Livingstone, D.N. (1992) *The Geographical Tradition: Episodes in the History of a Contested Enterprise*, Oxford: Blackwell.

Livingstone, D.N. and Harrison, R.T. (1981) "Immanuel Kant, subjectivism, and human geography: A preliminary investigation," *Transactions of the Institute of British Geographers*, New Series, 6.3: 359–74.

Locke, J. (1690/1967) *Two Treatises of Government*, New York: Cambridge University Press.

Lösch, A. (1938) "The nature of economic regions," *Southern Economic Journal*, 5.1: 71–78.

Louw, D.J. (2006) "The African concept of *ubuntu* and restorative justice," in D. Sullivan and L. Tifft (eds), *Handbook of Restorative Justice: A Global Perspective*, 161–73, New York: Routledge.

Luoma-aho, M. (2002a) "Europe as a living organism: Organicist symbolism and political subjectivity in the New Europe," Ph.D. dissertation, Department of Politics, University of Newcastle upon Tyne, UK.

—— (2002b) "Body of Europe and malignant nationalism: A pathology of the Balkans in European security discourse," *Geopolitics*, 7.3: 117–42.

Luxemburg, R. (1913/1951) *The Accumulation of Capital*, New Haven, CT: Yale University Press.

McCann, E.J. (2002) "The urban as an object of study in global cities literatures: Representational practices and conceptions of place and scale," in A. Herod and M.W. Wright (eds), *Geographies of Power: Placing Scale*, 61–84, Oxford: Blackwell.

—— (2004) "Urban political economy beyond the 'Global City'," *Urban Studies*, 41.12: 2315–33.

McCarthy, J. (2005) "Scale, sovereignty, and strategy in environmental governance," *Antipode*, 37.4: 731–53.

McCarty, H.H. (1940) *The Geographic Basis of American Economic Life*, New York: Harpers and Brothers.

McClintock, A. (1995) *Imperial Leather: Race, Gender, and Sexuality in the Colonial Contest*, New York: Routledge.

McDonald, J.R. (1966) "The region: Its conception, design, and limitations," *Annals of the Association of American Geographers*, 56.3: 516–28.

McDowell, L. (1983) "Towards an understanding of the gender division of urban space," *Environment and Planning D: Society and Space*, 1.1: 59–72.

—— (1993) "Space, place and gender relations, part I: Feminist empiricism and the geography of social relations," *Progress in Human Geography*, 17.2: 157–79.

McKinsey & Company (2005) *$118 Trillion and Counting: Taking Stock of the World's Capital Markets*.

Mac Laughlin, J. (2001) *Reimagining the Nation-State: The Contested Terrains of Nation-Building*, London: Pluto Press.

MacLeod, G. and Jones, M. (2001) "Renewing the geography of regions," *Environment and Planning D*, 16.9: 669–95.

—— (2007) "Territorial, scalar, networked, connected: In what sense a 'regional world'?," *Regional Studies*, 41.9: 1177–91.

McLuhan, M. (1964/2001) *Understanding Media: The Extensions of Man*, 2nd edn, New York: Routledge.

—— (1969/1997) "Playboy interview: 'Marshall McLuhan – A candid conversation with the high priest of popcult and metaphysician of media'," in E. McLuhan and F. Zingrone (eds), *Essential McLuhan*, 222–60, London: Routledge.

—— and Fiore, Q. (1967) *The Medium is the Massage*, New York: Bantam.

McMaster, R.B. and Sheppard, E. (2004) "Introduction: Scale and geographic inquiry," in E. Sheppard and R.B. McMaster (eds), *Scale and Geographic Inquiry: Nature, Society, and Method*, 1–22, Oxford: Blackwell.

Macrae, H. (2006) "Rescaling gender relations: The influence of European directives on the German gender regime," *Social Politics: International Studies in Gender, State and Society*, 13.4: 522–50.

Madden, J.F. (1981) "Why women work closer to home," *Urban Studies*, 18: 181–94.

Mahon, R. (2006) "Introduction: Gender and the politics of scale," *Social Politics: International Studies in Gender, State and Society*, 13.4: 457–61.

Mains, S.P. (2002) "Maintaining national identity at the border: Scale, masculinity, and the policing of immigration in Southern California," in A. Herod and M.W. Wright (eds), *Geographies of Power: Placing Scale*, 192–214, Oxford: Blackwell.

Mann, M. (1984) "The autonomous power of the state: Its origins, mechanisms and results," *European Journal of Sociology*, 25.2: 185–213.

Margulis, L. (1970) *Origin of Eukaryotic Cells: Evidence and Research Implications for a Theory of the Origin and Evolution of Microbial, Plant, and Animal Cells on the Precambrian Earth*, New Haven, CT: Yale University Press.

Markusen, A. (1996) "Sticky places in slippery space: A typology of industrial districts," *Economic Geography*, 72.3: 293–313.

Marston, S.A. (2000) "The social construction of scale," *Progress in Human Geography*, 24.2: 219–42.

—— (2002) "A long way from home: Domesticating the social construction of scale," in R. McMaster and E. Sheppard (eds), *Scale and Geographic Inquiry: Nature, Society and Method*, 170–91, Oxford: Blackwell.

Marston, S.A. and Smith, N. (2001) "States, scales and households: Limits to scale thinking? A response to Brenner," *Progress in Human Geography*, 25.4: 615–19.

Marston, S.A., Jones, J.P. III and Woodward, K. (2005) "Human geography without scale," *Transactions of the Institute of British Geographers*, New Series, 30.4: 416–32.

Marston, S.A., Woodward, K. and Jones, J.P. III (2007) "Flattening ontologies of globalization: The Nollywood case," *Globalizations*, 4.1: 45–63.

Martin, G.J. (1994) "In memoriam: Richard Hartshorne, 1899–1992," *Annals of the Association of American Geographers*, 84.3: 480–92.

—— (2005) *All Possible Worlds: A History of Geographical Ideas*, 4th edn, New York: Oxford University Press.

Marx, K. (1844/2007) *Economic and Philosophic Manuscripts of 1844*, Mineola, NY: Dover Publications.

—— (1858/1973) *Grundrisse: Foundations of the Critique of Political Economy*, London: Penguin.

—— (1867/1976) *Capital: A Critique of Political Economy, Volume One*, London: Penguin (1990 printing).

—— (1894/1981) *Capital: A Critique of Political Economy, Volume Three*, London: Penguin (1991 printing).

Marx, K. and Engels, F. (1845/1970) *The German Ideology (Part One)*, New York: International Publishers (1991 Printing).

—— (1848/1948) *The Communist Manifesto*, New York: International Publishers.

Massey, D. (1984) *Spatial Divisions of Labour: Social Structures and the Geography of Production*, London: Macmillan.

—— (1993) "Questions of locality," *Geography*, 78.2: 142–49.

—— (2004) "Geographies of responsibility," *Geografiska Annaler B*, 86.1: 5–18.

Masson, D. (2006) "Constructing scale/contesting scale: Women's movement and rescaling politics in Québec," *Social Politics: International Studies in Gender, State and Society*, 13.4: 462–86.

Mauss, M. (1936/1979) "The notion of body techniques," in M. Mauss, *Sociology and Psychology: Essays* (translated by B. Brewster), 97–105, London: Routledge.

May, J.A. (1970) *Kant's Concept of Geography and its Relation to Recent Geographical Thought*, Toronto: University of Toronto Press.

Meentemeyer, V. (1989) "Geographical perspectives of space, time, and scale," *Landscape Ecology*, 3.3/4: 163–73.

Meijers, E. (2007a) *Synergy in Polycentric Urban Regions: Complementarity, Organising Capacity and Critical Mass*, Amsterdam: IOS Press.

—— (2007b) "From central place to network model: Theory and evidence of a paradigm change," *Tijdschrift voor economische en sociale geografie*, 98.2: 245–59.

Meller, H.E. (1990) *Patrick Geddes: Social Evolutionist and City Planner*, New York: Routledge.

Meltzer, A. (1996) *Anarchism: Arguments For and Against*, San Francisco: A.K. Press.

Merchant, C. (1980) *The Death of Nature: Women, Ecology, and the Scientific Revolution*, San Francisco: Harper and Row.

Mercier, G. (1995) "La région et l'état selon Friedrich Ratzel et Paul Vidal de la Blache" ["The region and the state according to Friedrich Ratzel and Paul Vidal de la Blache"], *Annales de géographie*, 583: 211–35.

Merleau-Ponty, M. (1945/2002) *Phenomenology of Perception*, New York: Routledge.

Michael, K. and Masters, A. (2005) "Applications of human transponder implants in mobile commerce," Faculty of Informatics – Papers, University of Wollongong, Wollongong, New South Wales, Australia.

Mill, H.R., Herbertson, A.J., Freshfield, D., Oldham, Y., Ravenstein, E.G. and Mackinder, H. (1905) "The major natural regions: An essay in systematic geography: Discussion," *The Geographical Journal*, 25.3: 310–12.

Miller, B. (1997) "Political action and the geography of defense investment: Geographical scale and the representation of the Massachusetts Miracle," *Political Geography*, 16.2, 171–85.

Mills, L. (2006) "Maternal health policy and the politics of scale in Mexico," *Social Politics: International Studies in Gender, State and Society*, 13.4: 487–521.

Milward, A.S. (1992) *The European Rescue of the Nation-State*, Berkeley and Los Angeles: University of California Press.

Mitchell, D. (2000) *Cultural Geography: A Critical Introduction*, Oxford: Blackwell.

Mitchell, T. (1998) "Fixing the economy," *Cultural Studies* 12.1, 82–101.

Moore, A. (2008) "Rethinking scale as a geographical category: From analysis to practice," *Progress in Human Geography* 32.2, 203–25.

Moore, D.S. (1997) "Remapping resistance: 'Ground for struggle' and the politics of place," in S. Pile and M. Keith (eds), *Geographies of Resistance*, 87–106, London: Routledge.

Morgan, K. (2007) "The polycentric state: New spaces of empowerment and engagement?," *Regional Studies*, 41.9: 1237–51.

Morgan, W. (2000) "Queering international human rights law," in C.F. Stychin and D. Herman (eds), *Sexuality in the Legal Arena*, 208–25, London: Athlone.

Mosse, G.L. (1988) *Nationalism and Sexuality: Middle-Class Morality and Sexual Norms in Modern Europe*, Madison: University of Wisconsin Press.

Muir, J. (1916) *My First Summer in the Sierra*, Boston: Houghton Mifflin.

Mulgan, G.J. (1991) *Communication and Control: Networks and the New Economies of Communication*, New York: Guilford.

Mumford, L. (1961) *The City in History: Its Origins, its Transformations, and its Prospects*, New York: Harcourt, Brace and World.

Murdoch, J. (1995) "Actor-networks and the evolution of economic forms: Combining description and explanation in theories of regulation, flexible specialization, and networks," *Environment and Planning A*, 27.5: 731–57.

Murray, J. (2000) "Of nodes and networks: Bruges and the infrastructure of trade in fourteenth-century Europe," in P. Stabel, B. Blondé, and A. Greve (eds), *International Trade in the Low Countries (14th–16th Centuries): Merchants, Organisation, Infrastructure*, 1–14, Leuven: Garant.

Mussolini, B. (1932) "The doctrine of fascism." Available at www.worldfuture fund.org/wffmaster/Reading/Germany/mussolini.htm and copied from an official Fascist government publication of 1935, *Fascism: Doctrine and Institutions* (Benito Mussolini, Ardita Publishers, Rome, pages 7–42); last accessed December 12, 2009.

Naddaf, G. (2005) *The Greek Concept of Nature*, Albany, NY: SUNY Press.

Nagahara, Y. (2000) "Monsieur le capital and Madame la terre do their ghost-dance: Globalization and the nation-state," *The South Atlantic Quarterly*, 99.4, 929–61.

Nast, H.J. (2005) *Concubines and Power: Five Hundred Years in a Northern Nigerian Palace*, Minneapolis: University of Minnesota Press.

Neocleous, M. (2003) "The political economy of the dead: Marx's vampires," *History of Political Thought*, 24.4: 668–84.

Newman, D. and Paasi, A. (1998) "Fences and neighbours in the postmodern world: Boundary narratives in political geography," *Progress in Human Geography*, 22.2: 186–207.

Newton, I. (1687/1999) *The Principia: Mathematical Principles of Natural Philosophy*, Berkeley and Los Angeles: University of California Press.

Nicholson, L. (1994) "Interpreting gender," *Signs: Journal of Women in Culture and Society*, 20.1: 79–105.

Nir, D. (1987) "Regional geography considered from the systems' approach," *Geoforum*, 18.2: 187–202.

Norris, F. (1901/1967) *The Octopus: A Story of California*, Port Washington: Kennikat Press.

Nystuen, J.D. and Dacey, M.F. (1961) "A graph theory interpretation of nodal regions," *Papers and Proceedings of the Regional Science Association*, 7: 29–42.

Ogilvie, A.G. (ed.) (1928/1952) *Great Britain: Essays in Regional Geography*, Cambridge: Cambridge University Press.

Ohmae, K. (1995) *The End of the Nation State: The Rise of Regional Economies*, New York: McKinsey and Company.

—— (2005) *The Next Global Stage: Challenges and Opportunities in Our Borderless World,* Upper Saddle River, NJ: Wharton School Publishing.

Oikkonen, V. (2004) "Mad embodiments: Female corporeality and insanity in Janet Frame's *Faces in the Water* and Sylvia Plath's *The Bell Jar*," *Helsinki English Studies*, 3 (online publication; no page numbers).

Omi, M. and Winant, H. (1994) *Racial Formation in the United States: From the 1960s to the 1990s*, 2nd edn, New York: Routledge.

Osiander, A. (2001) "Sovereignty, international relations, and the Westphalian myth," *International Organization*, 55.2: 251–87.

Ó Tuathail, G. (1996) *Critical Geopolitics: The Politics of Writing Global Space*, Minneapolis: University of Minnesota Press.

Paasi, A. (1991) "Deconstructing regions: Notes on the scales of spatial life," *Environment and Planning A*, 23.2: 239–56.

Pacione, M. (2005) *Urban Geography: A Global Perspective*, 2nd edn, New York: Routledge.

Padgett, J.F. and Ansell, C.K. (1993) "Robust action and the rise of the Medici, 1400–1434," *The American Journal of Sociology*, 98.6: 1259–1319.

Papazian, T. (2008) "State at war, state in war: The Nagorno-Karabakh conflict and state-making in Armenia, 1991–95," *The Journal of Power Institutions in Post-Soviet Societies*, 8 (online publication; no page numbers).

Park, R., Burgess, E.W. and McKenzie, R.D. (1925) *The City*, Chicago: University of Chicago Press.

Parkin, F. (1979) *Marxism and Class Theory: A Bourgeois Critique*, New York: Columbia University Press.

Paterson, J.H. (1974) "Writing regional geography – Problems and prospects in the Anglo-American realm," *Progress in Geography*, 6: 1–26.

Peattie, R. (1935) "Response to Platt, R.S. (1935) Field approach to regions," *Annals of the Association of American Geographers*, 25.3: 172.

Peck, J. (2001) *Workfare States*, New York: Guilford.

Peet, R. (1985) "The social origins of environmental determinism," *Annals of the Association of American Geographers*, 75.3: 309–33.

—— (1998) *Modern Geographical Thought*, Oxford: Blackwell.

Penck, A. (1927) "Geography among the Earth sciences," *Proceedings of the American Philosophical Society*, 66: 621–44.

Pickles, J. (1985) *Phenomenology, Science, and Geography: Space and the Human Sciences*, Cambridge: Cambridge University Press.

Pickover, C.A. (2006) *The Möbius Strip: Dr. August Möbius's Marvelous Band in Mathematics, Games, Literature, Art, Technology, and Cosmology*, New York: Thunder's Mouth.

Pike, B. (1981) *The Image of the City in Modern Literature*, Princeton, NJ: Princeton University Press.

Pile, S. (1996) *The Body and the City: Psychoanalysis, Space and Subjectivity*, London: Routledge.

Plato (2000) *The Republic* (translated by B. Jowett), Mineola, NY: Courier Dover Publications.

Platt, R.S. (1935) "Field approach to regions," *Annals of the Association of American Geographers*, 25.3: 153–72.

Polanyi, K. (1957) *The Great Transformation*, Boston: Beacon Press.

Pollard, H.P. (1993) *Tariacuri's Legacy: The Prehispanic Tarascan State*, Norman: University of Oklahoma Press.

Pratt, M.L. (1992) *Imperial Eyes: Travel Writing and Transculturation*, New York: Routledge.

Pred, A. (1977) *City-Systems in Advanced Economies: Past Growth, Present Processes, and Future Development Options*, New York: Wiley.

—— (1984) "Place as historically contingent process: Structuration and the time-geography of becoming places," *Annals of the Association of American Geographers*, 74.2: 279–97.

Price-Chalita, P. (1994) "Spatial metaphor and the politics of empowerment: Mapping a place for feminism and postmodernism in geography?," *Antipode*, 26.3: 236–54.

Puar, J.K. (2007) *Terrorist Assemblages: Homonationalism in Queer Times*, Durham, NC: Duke University Press.

Pudup, M.B. (1988) "Arguments within regional geography," *Progress in Human Geography*, 12.3: 369–90.

Purcell, M. (2003) "Islands of practice and the Marston/Brenner debate: Toward a more synthetic critical human geography," *Progress in Human Geography*, 27.3: 317–32.

Radice, H. (1984) "The national economy – a Keynesian myth?," *Capital and Class*, 22: 111–40.

Randall, J.H. (1976) *The Making of the Modern Mind: A Survey of the Intellectual Background of the Present Age*, New York: Columbia University Press.

Ratzel, F. (1897) *Politische Geographie [Political Geography]*, Munich and Leipzig: R. Oldenbourg.

Reilly, W.J. (1931) *The Law of Retail Gravitation*, New York: Knickerbocker Press.

Redfield, R. (1941) *The Folk Culture of Yucatan*, Chicago: University of Chicago Press.

Renner, G.T. (1935) "The statistical approach to regions," *Annals of the Association of American Geographers*, 25.3: 137–52.

Rich, A. (1980) "Compulsory heterosexuality and lesbian existence," *Signs: Journal of Women in Culture and Society*, 5.4: 631–60.

Robertson, R. and Khondker, H.H. (1998) "Discourses of globalization: Preliminary considerations," *International Sociology*, 13.1: 25–40.

Rose, G. (1993) *Feminism and Geography: The Limits of Geographical Knowledge*, Minneapolis: University of Minnesota Press.

Rosenau, J.N. (1999) "Toward an ontology for global governance," in M. Hewson and T.J. Sinclair (eds), *Approaches to Global Governance Theory*, 287–301, Albany, NY: SUNY Press.

Rössler, M. (1989) "Applied geography and area research in Nazi society: Central place theory and planning, 1933–45," *Environment and Planning D: Society and Space*, 7.4: 419–31

Rothbard, M. (1972) "Exclusive interview with Murray Rothbard," *The New Banner: A Fortnightly Libertarian Journal*, February 25. Available at www.lewrockwell.com/rothbard/rothbard103.html; last accessed December 12, 2009.

Rousseau, J.-J. (1762/1935) *The Social Contract*, New York: Dutton.

Rowles, G.D. (1978) *Prisoners of Space? Exploring the Geographical Experience of Older People*, Boulder, CO: Westview Press.

Roxby, P.M. (1926) "The theory of natural regions," *Geographical Teacher*, 13: 376–82.

Ryan, S. (1996) *The Cartographic Eye: How Explorers Saw Australia*, New York: Cambridge University Press.

Sack, R.D. (1974) "Chorology and spatial analysis," *Annals of the Association of American Geographers*, 64.3: 439–52.

Saegert, S. (1980) "Masculine cities and feminine suburbs: Polarized ideas, contradictory realities," *Signs: Journal of Women in Culture and Society*, 5.S3: S96-S111.

Sahlins, P. (1989) *Boundaries: The Making of France and Spain in the Pyrenees*, Berkeley and Los Angeles: University of California Press.

Said, E.W. (1978) *Orientalism*, New York: Pantheon Books.

Sanguin, A.-L. (1993) *Vidal de la Blache: Un génie de la géographie* [*Vidal de la Blache: Geographic Genius*], Paris: Belin.

Sassen, S. (1991) *The Global City: New York, London, Tokyo*, Princeton, NJ: Princeton University Press.

—— (2003) "Globalization or denationalization?," *Review of International Political Economy*, 10.1: 1–22.

Sauer, C.O. (1924) "The survey method in geography and its objectives," *Annals of the Association of American Geographers*, 14.1: 17–33.

—— (1925) "The morphology of landscape," *University of California Publications in Geography*, 2.2: 19–53.

—— (1931) "Cultural geography," in E.R.A. Seligman and A. Johnson (eds), *Encyclopaedia of the Social Sciences*, Volume 6, 621–24, New York: Macmillan.

—— (1941) "Foreword to historical geography," *Annals of the Association of American Geographers*, 31.1: 1–24.

—— (1964/1987) "'Now this matter of cultural geography': Notes from Carl Sauer's last seminar at Berkeley," edited by J.J. Parsons, in M.S. Kenzer (ed.), *Carl O. Sauer – A Tribute*, 153–63, Corvallis: Oregon State University Press.

Saunders, P. (1986) *Social Theory and the Urban Question*, 2nd edn, New York: Holmes and Meier.

Sayer, A. (1984) *Method in Social Science: A Realist Approach*, London: Hutchinson.

—— (1989) "The 'new' regional geography and problems of narrative," *Environment and Planning D*, 7.3: 253–76.

—— (2004) "Seeking the geographies of power," *Economy and Society*, 33.2: 255–70.

Schaefer, F.K. (1953) "Exceptionalism in geography: A methodological examination," *Annals of the Association of American Geographers*, 43.3: 226–49.

Schein, R.H. (1997) "The place of landscape: A conceptual framework for interpreting an American scene," *Annals of the Association of American Geographers*, 87.4: 660–80.

Schlottmann, A. (2008) "Closed spaces: Can't live with them, can't live without them," *Environment and Planning D: Society and Space*, 26.5: 823–41.

Scholte, J.A. (1996) "The geography of collective identities in a globalizing world," *Review of International Political Economy*, 3.4: 565–607.

Scott, C.D. (2007) "Queer/Nation: From 'Nihon bungaku' to 'Nihongo bungaku'," paper presented at the Sixteenth Annual Meeting of the Association for Japanese Literary Studies, Princeton, NJ, November 2–4.

Semple, E.C. (1911) *Influences of Geographic Environment, On the Basis of Ratzel's System of Anthropo-geography*, New York: H. Holt and Co.

Serres, M. (1991/1995) "Second conversation: Method," in M. Serres with B. Latour, *Conversations on Science, Culture, and Time*, 43–76, Ann Arbor: University of Michigan Press.

Shands, K.W. (1999) *Embracing Space: Spatial Metaphors in Feminist Discourse*, Westport, CT: Greenwood.

Shaw, R.B. (1978) *A History of Railroad Accidents, Safety Precautions and Operating Practices*, Binghamton, NY: Vail-Ballou Press.

Sheller, M. and Urry, J. (2006) "The new mobilities paradigm," *Environment and Planning A*, 38.2: 207–26.

Shen, J. (2004) "Cross-border urban governance in Hong Kong: The role of state in a globalizing city-region," *The Professional Geographer*, 56.4: 530–43.

Sibley, D. (1998) "Sensations and spatial science: Gratification and anxiety in the production of ordered landscapes," *Environment and Planning A*, 30.2: 235–46.

Silber, I.F. (1995) "Space, fields, boundaries: The rise of spatial metaphors in contemporary sociological theory," *Social Research* 62.2: 323–55.

Silvey, R. (2004) "Power, difference and mobility: Feminist advances in migration studies," *Progress in Human Geography*, 28.4: 490–506.

Smajic, S. (2001) "The Ecstasy of Speed: A review of Paul Virilio, *A Landscape of Events*, Cambridge, MA: MIT Press.

——— 2000, *Postmodern Culture: An Electronic Journal of Interdisciplinary Criticism*, 12.1. Available at www.iath.virginia.edu/pmc/text-only/issue.901/12.1.r_smajic.txt; last accessed December 12, 2009.

Smith, C.A. (1976) "Regional economic systems: Linking geographical models and socioeconomic problems," in C.A. Smith (ed.), *Regional Analysis: Volume I – Economic Systems*, 3–63, New York: Academic Press.

Smith, M.P. (2001) *Transnational Urbanism: Locating Globalization*, Oxford: Blackwell.

Smith, N. (1981) "Degeneracy in theory and practice: Spatial interactionism and radical eclecticism," *Progress in Human Geography*, 5.1: 111–18.

——— (1984) "Deindustrialization and regionalization: Class alliance and class struggle," *Papers in Regional Science*, 54.1: 113–28.

——— (1984/1990) *Uneven Development: Nature, Capital and the Production of Space*, Oxford: Blackwell.

——— (1986) "On the necessity of uneven development," *International Journal of Urban and Regional Research*, 10.1: 87–104.

——— (1987) "Dangers of the empirical turn: Some comments on the CURS initiative," *Antipode*, 19.1: 59–68.

——— (1988) "The region is dead! Long live the region!," *Political Geography Quarterly*, 7: 141–52.

——— (1989a) "Rents, riots and redskins," *Portable Lower East Side*, 6: 1–36.

—— (1989b) "Geography as museum: Private history and conservative idealism in *The Nature of Geography*," in J.N. Entrikin and S.D. Brunn (eds), *Reflections of Richard Hartshorne's The Nature of Geography*, 91–120, Washington, DC: Occasional Publications of the Association of American Geographers.

—— (1992a) "Geography, difference and the politics of scale," in J. Doherty, E. Graham and M. Malek (eds), *Postmodernism and the Social Sciences*, 57–79, New York: St. Martin's Press.

—— (1992b) "Contours of a spatialized politics: Homeless Vehicles and the production of geographical scale," *Social Text*, 33: 54–81.

—— (1993) "Homeless/global: Scaling places," in J. Bird, B. Curtis, T. Putnam, G. Robertson and L. Tickner (eds), *Mapping the Futures: Local Cultures, Global Change*, 87–119, London: Routledge.

—— (1995) "Remaking scale: Competition and cooperation in prenational and postnational Europe," in H. Eskelinen and F. Snickars (eds), *Competitive European Peripheries*, 59–74, Berlin: Springer-Verlag.

—— (2000) "Scale," in R.J. Johnston, D. Gregory, G. Pratt and M. Watts (eds), *The Dictionary of Human Geography*, 724–27, Oxford: Blackwell.

—— (2003) *American Empire: America's Geographer and the Prelude to Globalization*, Berkeley and Los Angeles: University of California Press.

Smith, N. and Dennis, W. (1987) "The restructuring of geographical scale: Coalescence and fragmentation of the northern core region," *Economic Geography*, 63.2: 160–82.

Smith, N. and Katz, C. (1993) "Grounding metaphor: Towards a spatialized politics," in M. Keith and S. Pile (eds), *Place and the Politics of Identity*, 67–83, New York: Routledge.

Smith, R.G. (2003) "World city topologies," *Progress in Human Geography*, 27.5: 561–82.

Smith, W.D. (1980) "Friedrich Ratzel and the origins of lebensraum," *German Studies Review*, 3.1: 51–68.

Snow, D.A. and Benford, R.D. (1992) "Master frames and cycles of protest," in A.D. Morris and C.M. Mueller (eds), *Frontiers in Social Movement Theory*, 133–55, New Haven, CT: Yale University Press.

Søgaard, K., Blangsted, A.K., Herod, A. and Finsen, L. (2006) "Work design and the labouring body: Examining the impacts of work organisation on Danish cleaners' health," *Antipode*, 38.3: 579–602.

Soja, E. (1980) "The socio-spatial dialectic," *Annals of the Association of American Geographers*, 70.2: 207–25.

Solot, M. (1986) "Carl Sauer and cultural evolution," *Annals of the Association of American Geographers*, 76.4: 508–20.

Stalder, F. (2006) *Manuel Castells: The Theory of the Network Society*, Cambridge: Polity.

Steel, R.W. and Fisher, C.A. (eds) (1956) *Geographical Essays on British Tropical Lands*, London: George Philip.

Stern, D.I. (1992) "Do regions exist? Implications of synergetics for regional geography," *Environment and Planning A*, 24.10: 1431–48.

Stevens, A. (1939) "The natural geographical region," *Scottish Geographical Magazine*, 55.6: 305–17.

Stewart, S. (1984) *On Longing: Narratives of the Miniature, the Gigantic, the Souvenir, the Collection*, Baltimore: Johns Hopkins University Press.

Summerfield, M.A. (2005) "A tale of two scales, or the two geomorphologies," *Transactions of the Institute of British Geographers*, New Series, 30.4: 402–15.

Swyngedouw, E. (1996) "Reconstructing citizenship, the re-scaling of the state and the new authoritarianism: Closing the Belgian mines," *Urban Studies*, 33.8: 1499–1521.

—— (1997a) "Excluding the other: The production of scale and scaled politics," in R. Lee and J. Wills (eds), *Geographies of Economies*, 167–76, London: Arnold.

—— (1997b) "Neither global nor local: 'Glocalization' and the politics of scale," in K. Cox (ed.), *Spaces of Globalization: Reasserting the Power of the Local*, 137–66, New York: Guilford Press.

—— (2004) *Social Power and the Urbanization of Water: Flows of Power*, Oxford: Oxford University Press.

—— (2007) "Technonatural revolutions: The scalar politics of Franco's hydrosocial dream for Spain, 1939–75," *Transactions of the Institute of British Geographers*, New Series, 32.1: 9–28.

Swyngedouw, E. and Heynen, N.C. (2003) "Urban political ecology, justice and the politics of scale," *Antipode*, 35.5: 898–918.

Symanski, R. and Newman, J.L. (1973) "Formal, functional, and nodal regions: Three fallacies," *Professional Geographer*, 25.4: 350–52.

Tanner, N.M. (1981) *On Becoming Human: A Model of the Transition from Ape to Human and the Reconstruction of Early Human Social Life*, Cambridge: Cambridge University Press.

Tarrow, S. (2004) "From comparative historical analysis to 'local theory': The Italian city-state route to the modern state," *Theory and Society*, 33.3/4: 443–71.

Taylor, P.J. (1981) "Geographical scales within the world-economy approach," *Review*, 5: 3–11.

—— (1982) "A materialist framework for political geography," *Transactions of the Institute of British Geographers*, New Series, 7.1: 15–34.

—— (1987) "The paradox of geographical scale in Marx's politics," *Antipode*, 19.3, 287–306.

—— (1994) "The state as container: Territoriality in the modern world-system," *Progress in Human Geography*, 18.2: 151–62.

—— (1997) "Hierarchical tendencies amongst world cities: A global research proposal," *Cities*, 14.6: 323–32.

—— (2004) *World City Network: A Global Urban Analysis*, London: Routledge.

Teschke, B. (2003) *The Myth of 1648: Class, Geopolitics, and the Making of Modern International Relations*, New York: Verso.

Thomas, E. (2007) *Monumentality and the Roman Empire: Architecture in the Antonine Age*, Oxford: Oxford University Press.

Thornborrow, J. (1993) "Metaphors of security: A comparison of representation in defence discourse in post-Cold-War France and Britain," *Discourse & Society*, 4.1: 99–119.

Thrift, N. (1993) "For a new regional geography 3," *Progress in Human Geography*, 17.1: 92–100.

—— (2000) "Pandora's box? Cultural geographies of economies," in G. Clark, M. Feldmann and M. Gertler (eds), *The Oxford Handbook of Economic Geography*, 689–702, Oxford: Oxford University Press.

—— (2002) "A hyperactive world," in R.J. Johnston, P.J. Taylor and M.J. Watts (eds), *Geographies of Global Change: Remapping the World*, 2nd edn, 29–42, Oxford: Blackwell.

Tilly, C. (1985) "War making and state making as organized crime," in P.B. Evans, D. Rueschemeyer, and T. Skocpol (eds), *Bringing the State Back In*, 169–91, Cambridge: Cambridge University Press.

—— (1990) *Coercion, Capital, and European States, AD 990–1990*, Cambridge, MA: Blackwell.

Times [London] (1858) Friday, August 6, p. 8.

—— [London] (1866) No title, July 27, p. 9.

—— [London] (1866) No title, July 30, p. 8.

—— [London] (1866) Friday, September 21, p. 6.

Tönnies, F. (1887/ 2001) *Community and Civil Society*, Cambridge: Cambridge University Press.

Tower, W.S. (1908) "The human side of systematic geography," *Bulletin of the American Geographical Society*, 40.9: 522–30.

Tuan, Y.-F. (1982) *Segmented Worlds and Self: Group Life and Individual Consciousness*, Minneapolis: University of Minnesota Press.

Tunander, O. (2001) "Swedish-German geopolitics for a new century – Rudolf Kjellén's 'The State as a Living Organism'," *Review of International Studies*, 27.3: 451–63.

Unstead, J.F. (1916) "A synthetic method of determining geographical regions," *The Geographical Journal*, 48.3: 230–42.

Unwin, T. (2006) "100 years of British geography: The challenge of relevance," *Inforgeo*, 18/19: 103–26.

Urban Geography (2008) "Special issue on 'Chicago and Los Angeles: Paradigms, schools, archetypes, and the urban process'," *Urban Geography*, 29.2: 97–186.

Valentine, G. (1996) "(Re)negotiating the 'heterosexual street': Lesbian production of space," in N. Duncan (ed.), *BodySpace: Destabilizing Geographies of Gender and Sexuality*, 145–54, London: Routledge.

Van Cleef, E. (1930) "The geographical work of Alfred Hettner," *Geographical Review*, 20.2: 354–56.

Vance, J.E., Jr. (1970) *The Merchant's World: The Geography of Wholesaling*, Englewood Cliffs, NJ: Prentice Hall.

Vance, R.B. (1929) "The concept of the region," *Social Forces*, 8.2: 208–18.

Verne, J. (1872/2004) *Around the World in 80 Days*, Whitefish, MT: Kessinger Publishing.

Vidal de la Blache, P. (1903/1994) *Tableau de la géographie de la France* [*Picture of the Geography of France*], Paris: Table Ronde.

—— (1909) Review of "Les paysans de la Normandie orientale" ["The peasants of eastern Normandy"] by Jules Sion, *Annales de Géographie*, 18.98: 177–81.

—— (1913) "Des caractères distinctifs de la géographie" ["Some distinctive characteristics of geography"], *Annales de géographie*, 124: 289–99.

—— (1918/1926) *Principles of Human Geography*, New York: H. Holt and Co.

Virilio, P. (1993) "The third interval: A critical transition," in V.A. Conley and the Miami Theory Collective (Oxford, Ohio) (eds), *Rethinking Technologies*, 3–12, Minneapolis: University of Minnesota Press.

—— (1995a) *The Art of the Motor*, Minneapolis: University of Minnesota Press.

—— (1995b) "Speed and information: Cyberspace alarm!", originally published in French in *Le Monde Diplomatique*, August 1995. Available at www.ctheory.net/articles.aspx?id = 72; last accessed December 12, 2009.

—— (1997) *Open Sky*, New York: Verso.

Vlassopoulos, K. (2007) *Unthinking the Greek Polis: Ancient Greek History Beyond Eurocentrism*, Cambridge: Cambridge University Press.

von Böventer, E. (1963) "Toward a united theory of spatial economic structure," *Papers in Regional Science*, 10.1: 163–87.

Walker, D. and Walker, M. (2008) "Power, identity and the production of buffer villages in 'the second most remote region in all of Mexico'," *Antipode* 40.1: 155–77.

Walker, R.A. (1981) "A theory of suburbanization: Capitalism and the production of urban space in the United States," in M. Dear and A.J. Scott (eds), *Urbanization and Urban Planning in Capitalist Society*, 383–430, New York: Methuen.

Wallerstein, I. (1974) *The Modern World-System: Capitalist Agriculture and the Origins of the European World-Economy in the Sixteenth Century*, New York: Academic Press.

Ward, K. and Jonas, A.E.G. (2004) "Competitive city-regionalism as a politics of space: A critical reinterpretation of the new regionalism," *Environment and Planning A*, 36.12: 2119–39.

Warf, B. (2009) "From surfaces to networks," in B. Warf and S. Arias (eds), *The Spatial Turn: Interdisciplinary Perspectives*, 59–76, Abingdon, UK: Routledge.

Warwick, K. (2004) *I, Cyborg*, Urbana: University of Illinois Press.

Waterman, P. (1993) "Internationalism is dead! Long live global solidarity?," in J. Brecher, J. Brown Childs, and J. Cutler (eds), *Global Visions: Beyond the New World Order*, 257–61, Boston: South End Press.

Watson, J.W. (1983) "The soul of geography," *Transactions of the Institute of British Geographers*, New Series, 8.4: 385–99.

Weber, M. (1919/2004) *The Vocation Lectures*, Indianapolis, IN: Hackett Publishing.

—— (1921/1958) *The City*, Glencoe, IL: Free Press.

—— (1925/1978) *Economy and Society: An Outline of Interpretive Sociology (Volume II)*, Berkeley and Los Angeles: University of California Press.

Weber, R.E. (1992) "Seward's other folly: America's first encrypted cable," *Studies in Intelligence* 36.5, 105–9.

Weibe, S. (2009) "Producing bodies and borders: A review of immigrant medical examinations in Canada," *Surveillance & Society*, 6.2: 128–41.

Wells, H.G. (1898/2006) *The War of the Worlds*, Teddington, UK: Echo Library.

White, C.H.F. (1942) "Laying a submarine cable," *The Geographical Magazine*, January, 142–45.

Whittlesey, D. (1954) "The regional concept and the regional method," in P.E. James and C.F. Clarence (eds), *American Geography: Inventory and Prospect*, 19–68, Syracuse, NY: Syracuse University Press.

Wiebe, R. (1967) *The Search For Order 1877–1920*, New York: Hill and Wang.

Wills, J. (1998) "Taking on the CosmoCorps? Experiments in transnational labor organization," *Economic Geography*, 74: 111–30.

Wirth, L. (1938) "Urbanism as a way of life," *American Journal of Sociology*, 44.1: 1–24.

Wishart, D. (2004) "Period and region," *Progress in Human Geography*, 28.3: 305–19.

Witt, P.N., Scarboro, M.B., Daniels, R., Peakall, D.B. and Gause, R.L. (1977) "Spider web-building in outer space: Evaluation of records from the Skylab spider experiment," *American Journal of Arachnology*, 4:115–24.

Wolf, M. (2001) "Will the nation-state survive globalization?," *Foreign Affairs* 80.1: 178–90.

Wolfart, P.D. (2008) "Mapping the early modern state: The work of Ignaz Ambros Amman, 1782–1812," *Journal of Historical Geography*, 34.1: 1–23.

Woofter, T.J. (1934) "Subregions of the Southeast," *Social Forces*, 13.1: 43–50.

Wright, M.W. (2001) "Desire and the prosthetics of supervision: A case of maquiladora flexibility," *Cultural Anthropology*, 16.3: 354–73.

Wriston, W.B. (1992) *The Twilight of Sovereignty: How the Information Revolution Is Transforming Our World*, New York: Charles Scribner's Sons.

Zelinsky, W. (1973) *The Cultural Geography of the United States*, Englewood Cliffs, NJ: Prentice-Hall.

Zipf, G.K. (1946) "The P_1P_2D hypothesis: On the intercity movement of persons," *American Sociological Review*, 11.6: 677–86.

Zobler, L. (1955) "Statistical testing of regional boundaries," *Annals of the Association of American Geographers*, 47.1: 83–95.

INDEX

Printed in the United States
by Baker & Taylor Publisher Services